Japan's Population Implosion

Yoichi Funabashi
Editor

Japan's Population Implosion

The 50 Million Shock

palgrave
macmillan

Editor
Yoichi Funabashi
Asia Pacific Initiative
Rebuild Japan Initiative Foundation
Tokyo, Japan

ISBN 978-981-10-4982-8   ISBN 978-981-10-4983-5 (eBook)
https://doi.org/10.1007/978-981-10-4983-5

Library of Congress Control Number: 2017951397

Cover illustration: Design by Jenny Vong

Printed on acid-free paper

This Palgrave Macmillan imprint is published by Springer Nature
The registered company is Springer Nature Singapore Pte Ltd.
The registered company address is: 152 Beach Road, #21-01/04 Gateway East, Singapore 189721, Singapore

Preface

Neither the government nor the National Diet has fully confronted the crisis of Japan's population problem. While politicians and government officials dawdle, the population has moved into absolute decline. But it is not only politicians and government officials whom we should hold responsible. We, the Japanese people, have also failed to pay sufficient attention to this question. The authors of this volume and I were all very slow to notice the seriousness of this grave problem. It was, in fact, at the time of the 2011 Great East Japan Earthquake that the facts about aging population and regional exhaustion, along with the resultant weakening of Japan's basic resilience as a country, were brought most clearly to the forefront. When we went to Fukushima Prefecture for fieldwork during the Rebuild Japan Initiative Foundation (RJIF)'s first project, the Independent Investigation Commission on the Fukushima Nuclear Accident, I personally became acutely aware that we must all address the population issue.

This awareness that the issue had turned critical was taken up later in RJIF's project reports: *Japan in Peril: 9 crisis scenarios (2013)* and *Examining Japan's Lost Decades (2015)*. However, in the process of this volume we decided to focus all our resources on the population question and follow the model of the Fukushima Investigation, to once again lead a private sector inquiry into the question from multiple perspectives, to examine the policy failures, and to make policy recommendations. To move forward with administrative reform, the government set up in 1981 a Second Ad Hoc Commission on Administrative Reform chaired by the former chairman of the Keidanren (Japan Business Federation), Doko Toshio. Doko pursued this initiative energetically, declaring at the Commission's outset, "We will squarely face and examine the issues. We will not be impeded to find whatever problems exist regardless of where they are." Wishing to uphold this spirit, we launched the Private Population Ad Hoc Commission in April 2014.

We were fortunate to be able to bring together the finest team, including Onishi Takashi (President, Toyohashi University of Technology/ President, The Science Council of Japan), Katsu Eijiro (President, COO and Representative Director, Internet Initiative Japan Inc.), Kigawa Makoto (Representative Director and Chairman, Yamato Holdings Co., Ltd.), Kobayashi Hikaru (Professor, The Graduate School of Media and Governance, Keio University), Shirakawa Masaaki (Professor, The Graduate School of International Politics, Economics and Communications, Aoyama Gakuin University), Deguchi Haruaki (Representative Director, Chairman and CEO, Lifenet Insurance Company), and Momoi Mariko (Vice President, International University of Health and Welfare). Professor Onishi willingly undertook the position of chair, and he noted that our methods were exactly how Japan needed to tackle this issue.

In the previous government administration Professor Onishi had chaired the Ministry of Land, Infrastructure, Transport & Tourism's Committee on the Long-Term Outlook for National Land. Other committee members were also convinced of the importance of this issue and all agreed with the spirit of this project. We had to create a working group to support the Commission.

In making policy recommendations, RJIF's stance is to always bear in mind how to overcome political constraints through a rigorous examination of previous policy initiatives and failures. We fully investigate, conduct interviews directly with experts, mainly parties actually involved in the policy debates, and on this basis aim at identifying realistic policy alternatives. In the year's work carried out between April 2014 and March 2015, we conducted 10 meetings with experts, 17 working group meetings, and about 21 in-depth interviews. On March 5, 2015, we invited Ishiba Shigeru, Minister for Regional Regeneration, to discuss our draft policy recommendations and many of those suggestions are directly reflected in this report. A list of Commission and working group members as well as interviewees is included elsewhere in this volume, but I would like to express my sincere gratitude to all the Commission members beginning with Chairman Onishi, as well as to the policy makers and experts who served as members of the working group, and all the interviewees including first and foremost Minister Ishiba.

The report was compiled under the aegis of the Commission. The Commission indicated the larger direction and policies, after which the working group composed a draft report. The Commission then commented on this and those comments were woven into a subsequent draft, which the Commission proofed and boiled down to the essentials. The finished first draft was edited by Tamura Eiji and turned into the final report. I would like to offer my sincere thanks to Tamura-san.

Building on the RJIF's *Japan in Peril: 9 crisis scenarios (2013)*, this report was also published by the Shinchosha Publishing Co., Ltd. I offer my deepest gratitude to its two directors, Ito Yukihito (Public Relations and Publicity Division) and Hara Hirosuke (Non-fiction Department).

March 23, 2015
Tokyo, Japan

Funabashi Yoichi
Chairman, Rebuild Japan
Initiative Foundation

Book Description

A nation's population is the beating heart that shapes a country, the source of its energy and soul. A country's size governs its economy, people's living standards, the national well-being, and a nation's power. Should Japan's current birth rate remain unchanged, the population, which stood at nearly 50 million people a century ago, will shrink once again to this size by the end of the twenty-first century. More significantly, Japan will not be a country overflowing with youthful exuberance as during the Meiji Era (1868–1912). It will, instead, turn into an old country where the elderly account for 40% of the population. If such a scenario occurs, Japan's vitality will deflate, the nation will lose confidence, and the society will fall into a state of despair.

The book examines the population crisis from a variety of perspectives, including population and land policies, public finance, and crisis management, but also offers policy proposals. Population decline is Japan's greatest challenge and cannot be postponed.

Contents

Contributors

Funabashi Yoichi is Chairman of the Asia Pacific Initiative and former Editor-in-Chief and Columnist for the *Asahi Shimbun*. He is the first Japanese recipient of the Shorenstein Journalism Award given by Stanford University's Walter H. Shorenstein Asia-Pacific Center and won the Japan Press Award, Japan's "Pulitzer Prize", and others.

Private Ad Hoc Commission Members

Deguchi Haruaki is Representative Director, Chairman and CEO of Lifenet Insurance Company, an innovative insurance service company he founded in 2008.

Katsu Eijiro is President and COO, and Representative Director of Internet Initiative Japan Inc., serving since 2013. He worked at the Ministry of Finance in Japan and served as Vice Minister from 2010 to 2012.

Kigawa Makoto is Representative Director and Chairman of Yamato Holdings Co., Ltd.

Kobayashi Hikaru is a Professor at the Graduate School of Media and Governance, Keio University.

Momoi Mariko is Vice President of the International University of Health and Welfare. She is also President of the International University of Health and Welfare Hospital.

Onishi Takashi is President of Toyohashi University of Technology and President of the Science Council of Japan (SCJ). He is also Professor Emeritus at the University of Tokyo.

Shirakawa Masaaki is a Professor at the Graduate School of International Politics, Economics and Communications, Aoyama Gakuin University, and former Governor of the Bank of Japan.

Working Group Members

Akiyama Yuki is an Assistant Professor at the Center for Spatial Information Science, the University of Tokyo. He specializes in Spatial Information Science with a focus on the development of Micro Geodata (MGD) and its application for various regional problems.

Arai Junji is a Senior Manager in the Product Strategies Department of LAWSON, Inc., one of the largest convenience store franchise companies in Japan.

Igarashi Chikako is President of General Incorporated Association Hokkaido Intellect Tank. Her major areas of research are in welfare-related topics, such as elderly care management and community welfare.

Inagawa Hidekazu is a Professor at the Faculty of Sociology, Toyo University. He specializes in social security law and social welfare policy. He previously served as a government official of the Ministry of Health, Labour and Welfare.

Kamiya Kenichi is a Researcher at the Medical Care Research Department, General Incorporated Association Hokkaido Intellect Tank.

Kato Sota is a Professor at the International University of Japan. He is concurrently Executive Director of Policy Research and a Senior Fellow at the Tokyo Foundation.

Matsuda Kiyoto is a Partner at Unison Capital, Inc. Prior to joining the company, he served as the Managing Executive Officer at Mizuho Corporate Bank and Deputy President of Mizuho Securities.

Oguro Kazumasa is a Professor at the Faculty of Economics, Hosei University, and Director of the Kajima Institute of International Peace. He served as Senior Researcher at the Policy Research Institute, Ministry of Finance.

Otake Hiroshi is a Professor and Chairman at the Department of Anesthesiology and Critical Care Medicine, Showa University School of Medicine. He is also a Vice Director of Showa University Hospital.

Nagao Takashi is Manager of the Division of Business Strategy, Yamato Transport Co., Ltd.

Seta Fumihiko is an Associate Professor at the Department of Urban Engineering of the Graduate School of Engineering, the University of Tokyo. His research themes include land and urban planning, regional development, and global city studies.

Shibasaki Ryosuke is a Professor at the Center for Spatial Information Science, the University of Tokyo. He served as Director of the center from 2005 to 2010. He also served as President of GIS Association of Japan.

Sunahara Yosuke is a Professor of Public Administration at the Graduate School of Law, Kobe University. He specializes in public administration and local government.

Umeyama Goro is a consultant specializing in risk and crisis management. He served as a working group leader of the Independent Investigation Commission on the Fukushima Nuclear Accident (Part 2) for the Rebuild Japan Initiative Foundation.

Project Staff

Capistrano R. Andrew Research Assistant, Asia Pacific Initiative.

Kushner Barak Reader in Japanese History (Associate Professor), Department of East Asian Studies/Faculty of Asian and Middle Eastern Studies, University of Cambridge.

Mizukoshi Kaeko Assistant, Asia Pacific Initiative.

Otomo Miyuki Public Affairs Officer and Fellow, Asia Pacific Initiative.

Sigyo Chihei Assistant, Asia Pacific Initiative.

Spinks Wendy Translator.

Interviewees

*Titles and institutional affiliation at time of interview

Atoh Makoto President, Health, Labour and Welfare Statistics Association.

Fujii Ken Deputy Director, Kanto Regional Development Bureau, Ministry of Land, Infrastructure, Transport and Tourism.

Iguchi Yasushi Professor, School of Economics, Kwansei Gakuin University.

Iwabuchi Katsuyoshi Professor, Tohoku Fukushi University (Head, Next Generation Educational Center); Visiting Senior Staff Writer, Sankei Shimbun.

Kawakami Yukio Special Director for Research, Urban Research Institute Corporation.

Kitazawa Hiroyuki Deputy Head, Hatoyama Newtown Resident's Association.

Kohno Shigemi Professor Emeritus, Reitaku University.

Komine Takao Professor, Graduate School of Regional Policy Design, Hosei University.

Kunii Hideo President, The Shonai Bank, Ltd.

Makino Mitsuo Mayor, Iida City, Nagano Prefecture.

Mizoguchi Hidetsugu Managing Director and Corporate Planning Division Manager, Tokai Holdings Corporation.

Mori Masashi Mayor, Toyama City, Toyama Prefecture.

Ochiai Emiko Professor of Sociology, Kyoto University; Member of Science Council of Japan.

Oizumi Keiichiro Senior Economist, Economics Department, The Japan Research Institute, Limited.

Sanada Takamitsu Deputy General Manager, Economic Affairs Research Group, Sales Planning Division, Fukuoka Financial Group, Inc.

Suzuki Toru Director of Department, Department of Population Structure Research, National Institute of Population and Social Security Research.

Takahashi Tai Professor and Field Chief, Health Service Management Field, Graduate School, International University of Health and Welfare.

Tsuji Takuya Professor, Graduate School of Law, Hitotsubashi University.

Ujihara Rieko Chief, Multicultural Section, Gender Equality Division, Citizen Partnership Environment Department, Iida City, Nagano Prefecture.

Utsumi Yoshio President and CEO, Data Horizon Co., Ltd.

Yajima Hiroshi Director, Welfare Department, Arakawa Ward Office, Tokyo.

Yamashige Shinji Associate Professor, Graduate School of Economics, and School of International and Public Policy, Hitotsubashi University.

Yokota Koji Director, Managing Executive Officer, The Bank of Fukuoka.

Abbreviations

ANRE	Agency for Natural Resources and Energy
APU	Ritsumeikan Asia Pacific University
CSIS	Center for Spatial Information Science, the University of Tokyo
DID	Demographically intense district
FY	Fiscal year
GDP	Gross domestic product
GNI	Gross national income
HIT	General Incorporated Association Hokkaido Intellect Tank
ILO	International Labour Organization
IPSS	National Institute of Population and Social Security Research
JDP	Japan Democratic Party
JFA	Japan Franchise Association
LDP	Liberal Democratic Party of Japan
METI	Ministry of Economy, Trade and Industry
MEXT	Ministry of Education, Culture, Sports, Science and Technology
MHLW	Ministry of Health, Labour and Welfare
MHW	Ministry of Health and Welfare
MIC	Ministry of Internal Affairs and Communications
MLIT	Ministry of Land, Infrastructure, Transport and Tourism
MOF	Ministry of Finance Japan
MOFA	Ministry of Foreign Affairs of Japan
MOJ	Ministry of Justice
NPO	Non-Profit Organization
OECD	Organisation for Economic Co-operation and Development

TEPCO	Tokyo Electric Power Company
TFP	Total factor productivity
TFR	Total fertility rate

List of Figures

CHAPTER 1

Introduction

Inagawa Hidekazu

The rapid population decline now taking place before our eyes in Japan stands as the country's greatest challenge since the start of the Meiji Restoration in 1868. A country's citizenry is the very basis of the state and shapes the health of the nation. Fluctuations in the population govern the size of the economy, people's living standards, as well as the state of the nation and its power. Should the current birth rate remain unchanged, Japan's population, which stood at nearly 50 million people 100 years ago, will shrink once again to 50 million by the end of this century. Moreover, this will not be a country overflowing with youth as during the Meiji Era (1868–1912). Japan will instead be an old country where the elderly account for 40% of that 50 million, according to the National Institute of Population and Social Security Research (IPSS).[1] If this occurs, there is a strong chance that Japan's social vitality and the vital forces of the state will be debilitated, jealous political rivalries will be unleashed, and people will lose confidence and become trapped in a spiral of pessimism and nihilism. No longer will Japan be able to maintain the culture and traditions it has nurtured over the centuries.

This must not happen.

It must not be allowed to happen.

Inagawa. H (✉)
Faculty of Sociology, Toyo University, Tokyo, Japan

Y. Funabashi (ed.), *Japan's Population Implosion*,
https://doi.org/10.1007/978-981-10-4983-5_1

To avoid such a calamity, Japan needs to devise an effective population policy. Such a policy needs both ***mitigation*** measures for staging a recovery in the depopulation process as well as ***adaptation*** measures for maintaining and improving living standards. A broad national consensus that spans the generations in support of this policy and the sustained political will to achieve it over the long term are required. To create a steadfast population policy both resolve and a broad-ranging consensus are necessary and can together be thought of as Japan's "Second Constitution." Creating a Second Constitution is the greatest responsibility of people living today. With an eye toward the future, as human beings, we can all live for the next generation.

1.1 No Time to Lose

Establishing a population policy must start first by verifying the failure of previous policies (or, more accurately, examining the lack of a proper population policy). This is necessary because to merely follow the same approach will only lead to failure once again.

The Japanese government first raised the issue of a declining population in the 1992 *White Paper on the National Lifestyle*, using the term "declining birth rate."[2] Until 1972, the government had subscribed to a theory of overpopulation. However, the demographic projections of the Ministry of Health and Welfare (now the Ministry of Health, Labour and Welfare [MHLW]) continued to publish optimistic scenarios for a recovery in the birth rate in its demographic projections. They expected to offset declines with population increases from the children of the baby boomers. This rosy situational awareness was the first stumbling block. Although as a matter of course the government should bear responsibility for this overly optimistic outlook, the general public also underestimated the economic, political and social consequences of a declining population. And although the government showed signs of comprehensively addressing the unprecedented low birth rate with its 2003 enactment of the Basic Act for Measures to Cope with Society with a Declining Birth Rate, its impact on reversing population decline was minimal. As a consequence, after peaking in 1995, Japan's working-age population began to decrease, and the total population entered a phase of absolute decline after peaking in 2008.

In retrospect, the 20 years from 1995 to 2015, when the children of the baby boomers reached childbearing age, was a critical period. This was the era that had an irreplaceable opportunity to halt the population decline. And Japan missed it. During this time, the deterioration of the economy due to the collapse of the bubble economy and financial bankruptcies, coupled with greater flexibility in the employment base (i.e., the rise of non-regular employment) hit young people who should have been marrying and having children. When the dust had settled, the "late marriage" phenomenon, where there was an expectation that people would eventually get married, had transformed into a situation where not getting wed became the new norm. At the same time, the issue of an aging society was already the main social welfare priority, focusing on the problem of pensions, health care, and elder care, so that the declining birth rate problem took a back seat. The government, business community, and labor market also prioritized employment matters and the livelihoods of the elderly rather than focusing on the plight of young people and failed to make a committed effort to provide effective youth support.

More recently, the Abe administration has demonstrated its commitment to population issues, declaring, "we will look into policies that diverge from previous declining birth rate policies" (in its June 2014 cabinet decision on Basic Policies for Economic and Fiscal Management and Reform 2014), as well as establishing a town-people-work creation headquarters in the Cabinet Secretariat. Nevertheless, an effective population policy outline backed by budgetary resources is yet to appear.[3] One postponement after another has actually hastened the population decline in Japan and if the current birth rate remains unchanged, eventually there will be no Japanese left.

There is no time to lose. Another frightening aspect of this problem is that it does not end with the population. An aging and declining population, especially with a drop in the working-age population, depresses overall economic growth. Alongside Germany, Japan's working-age population per capita gross domestic product (GDP) growth rate is the highest of the Organisation for Economic Co-operation and Development (OECD) nations, but the per capita GDP growth rate has remained around the OECD average. This is due to the drop in the working-age population, which is sharper than the overall population decline. It is estimated that Japan will fall into low growth averaging approximately negative 0.1% after 2040 if the population decline remains unaddressed and a state of productivity stagnation also continues.[4] If the decline becomes steeper, it will lead to negative growth and eventually this will cause a collapse in the infrastructure of daily life—where people will be unable to conveniently access the necessary goods and services to live. In fact, this negative chain reaction has already begun.

Of Japan's numerous failures over the past 30 years—neglecting the economic bubble, postponing dealing with nonperforming loans, economic stagnation, the Fukushima nuclear accident—the failure of crafting an effective population policy is perhaps the greatest failure of all. But why did it fail?

It seems Japan was not sufficiently aware of the policy peculiarity involved in the population issue, which is that various efforts tried to be effective without any long-term or comprehensive approach.

An aging society is defined as one where the proportion of those aged 65 and older is higher than 7%, a situation that had already occurred in Japan from 1970, the year of the Osaka Expo. On the heels of this event, the decline in the birth rate began. The birth rate in Japan has been falling sharply since the late 1970s, meaning that a situation where the population size can be maintained in the long term (population replacement level is currently 2.07) has already been unsustainable for 40 years. Like gravity, the accumulated pressure of this buildup is on the verge of crushing Japan's population. Even if a nation tries to stave off a population decline, it would require a decades-long approach. And even if workers from abroad, or foreign workers, are introduced in a planned manner, it will take decades for people to become rooted in the new society as residents. The gestation period for an effective population policy is frighteningly long. Nations cannot, therefore, suddenly aim for the summit when responding to population pressures. Additionally, when dealing with an already accelerating population decline, policies are required on the one hand to correct for (mitigate) adverse effects, and on the other hand, to maintain a sustainable socio-economy in anticipation of the inevitable population decline.

Being forced to deploy this two-pronged strategy is not without its difficulties. Moreover, since population policy effects emerge only gradually it is difficult for political leaders to earn the political capital to implement, such policy achievements while in office. In short, no field is trickier to tackle politically than demographic issues. We need to be aware of these obstacles and when making recommendations on population policy the country needs to bear in mind the involved political difficulty enmeshed in such debates.

1.2 Four "Negative Chain Reactions"

Most Japanese were not fully aware of another factor: the speed of future population decline and the numerous negative chain reactions that accompany it. One major factor is the failure to create an effective national land policy in light of demographic trends. The government discussed a national land policy oriented toward decentralization, but in fact promoted the selection and concentration of regional population and economies, failing to pursue a vision of population relocation necessary for revitalizing the hinterland. This failure on national land policy mirrors those in forming population policy.

1.2.1 Risk that Population Decline Will Push Down Economic Growth and Shatter Public Finances

Japan's economic growth rate has been gradually declining over the past two decades, with average real growth rate languishing around 1%. Although sluggish growth in the first half of this period was primarily due to the impact of the collapse of the bubble economy, low growth in the second half has been affected by the declining birth rate and an aging population as seen in the labor force population peak in 2007. Currently in Japan, the aging of the population is taking place at a much faster pace than the economic growth rate. A rapid advance in the aging and decline of a population weakens income formation capacity in the economy, possibly causing chronic economic stagnation. Elsewhere, rapid aging sharply increases social security costs, leading to a persistent budget deficit. Japan's public debt to GDP ratio currently stands at more than 200%. This level is already higher than the prewar peak that led to runaway inflation immediately after the country's defeat in World War II, and is akin to a time bomb that may destroy Japan's finances and economy.

In his best-selling book, *Capital in the Twenty-First Century*, the French economist Thomas Piketty points out that it took the UK over a century to reduce public debt, which amounted to 200% of GDP during the Napoleonic Wars from the late eighteenth century to the early nineteenth century, down to 30%, and that France, once bankrupt after the French Revolution and whose public debt had risen to 70–80% of the GDP after the Franco–Prussian War, finally managed to curb its public debt after the World War I (Piketty 2014). In Japan's case, the aging and declining population is already an expansion factor similar to these "war budgets" that will only strengthen expansionary pressure in future decades. This makes improving public debt a much more difficult challenge.

1.2.2 Risk that Population Decline Will Disrupt the Infrastructure of Daily Life

Sharp population decline is rapidly raising the number of small municipalities. The smaller a municipality becomes, the harder it will be to maintain the life infrastructure that comprises the minimum livelihood base for people in the region. Water, power, roads, hospitals, nursing homes, postal services, schools, shopping malls, and convenience stores, all the services built up over a long period of time since the postwar era can no longer be adequately provided. The impact of population decline is already at an intensity referred to as "regional extinction," which is most pronounced in Japan's rural areas. There, regional cities with a population between 100,000 and 200,000 can no longer serve as a "population dam," preventing a flood of people flowing from rural to urban areas.

Despite the reversal of population trends from increasing to decreasing, and with strained public finances, many local governments still insist on adhering to the mentality of Japan's high-growth era, trying to provide services independently. However, this kind of overstretching will no longer be viable when the population drops. Public services, such as maintaining hospitals and high schools, become more difficult, and it becomes even more challenging to remain a self-sustaining city. As for private services, many municipalities are now forced to take over when supermarkets and gas stations withdraw.

Women living in regional areas between the ages of 20 and 39 who cannot find places of employment gravitate as if sucked into a vortex toward a small number of metropolises, led by Tokyo. This population shift has also accelerated the decline in rural populations. However, the collapse of the infrastructure of daily life will eventually extend to Greater Tokyo, and already the population of young people in commuter suburbs in the metropolitan area has started to decline. If this trend continues, in one or two decades certain municipalities in Saitama and Chiba Prefectures, and even places in Tokyo proper such as Adachi Ward, are predicted to reach levels of depopulation similar to Aomori Prefecture today. If life infrastructure in the metropolitan area were to collapse, the impact—such as worsening standards of hygiene and security—might make the regional situation pale in comparison. A collapse of life infrastructure would mean that the government could no longer guarantee people the "minimum standards of wholesome and cultural living" as laid down in Article 25 of the Japanese Constitution. Should this occur, the government would cease to have any meaning for people.

1.2.3 Risk that Population Decline Will Intensify Intergenerational Conflict

The existing social security system is based on the premise that all people are equal and bear the burden to support social security pensions, medical care, and elder care, eventually becoming beneficiaries themselves. The balance of benefits and burdens in this system consists of a social contract between the generations: social security costs incurred when young will generally be recouped as social security services when older. Rapid depopulation is likely to shatter such an idyllic vision of intergenerational fairness. Burgeoning social security costs from rapid aging and population decline will force an undue burden onto the working generation, especially the young. Moreover, as depopulation continues there is a danger that the imbalance will become structural with the advent of a so-called silver democracy, a politically dominant class comprised of the more numerous elderly, further exacerbating intergenerational conflict.

1.2.4 Risk that the State of the Nation and Its Power Will Be Debilitated, Lowering Japan's International Status

Rapid depopulation weakens the nation and its power. It lowers economic growth potential and discourages social vitality. People determine their commitment and willingness to build a future in concert with others not according to current conditions but based on their future outlook. Only if there is a prospect of growth will people take risks and gamble on opportunities. It is through friendly competition with others, and at the same time through cooperation and working together, that discovery, innovation, and creativity are realized.

As the population shrinks, people may be confined to small, homogenous groups who criticize each other's failures and will thus be unable to make decisions informed by a global perspective or to argue for new solutions. Additionally, there is a danger that a generous and freewheeling social spirit will disappear. Depopulation could lead to inequality and a concentration of hereditary wealth thorough inheritance and financial transfers to a small number of offspring, and such disparity could have a deep negative impact on economic growth.[5]

This would harm Japan's international image holding the world's "respect" as a member of the international community of advanced nations. Expectations for Japan as a "problem frontrunner" to innovate and implement cutting-edge approaches would also disappear. Not only would the country lose the merits that the size of its GDP and population gained for it in international politics, but by losing what it can best provide to the world, Japan's global presence would potentially diminish even further.

1.3 A Conflict of Values

All of these above risks are intricately interconnected, complicating any solution to the problem. For example, population decline forces the redrawing of political constituencies to address voter discrepancy, and young people in the districts to be redrawn may grow so anxious that they abandon the future of their areas, further accelerating the exodus of youth from the outlying areas. We may even see rural revolts throughout the country. One of the greatest challenges for creating a viable population policy is the unavoidable clash of diverse values. In Japan, political showdowns over values are the exception. However, the population issue takes place within an environment where values sharply conflict and this is a source of social and political tension. The following themes, for example, are likely examples of such value conflicts.

1.3.1 Which Should Take Precedence: Economic Superpower or Per Capita Income?

What should be the goal of a population policy? Maintaining an absolute population level or maintaining per capita national income?

The maintaining-population faction claims if Japan does not stay above a certain level of population, the state of the nation and its power will be whittled away. Even maintaining national security will be difficult.

On the other hand, the per-capita-income faction would refute the assumption that a drop in the population is necessarily bad. Would it not be enough if people were able to lead richer, more agreeable lives as individuals? Rather than trying to desperately stave off a population decline, perhaps Japan should set its sights on a social and national vision based on the premise of slower growth and a declining population?

The heart of the debate is the goal of a population policy. To simplify the question: this is a choice between the doctrine espousing a Great Japan (maintaining a population of 100 million people regarded as the minimum size for a major power) and one advocating for a Small Japan.

Regardless, we believe it is necessary to maintain a certain population level. If no action is taken, nothing in the end will be possible and Japan's population will dissipate with towns, villages, and peripheral islands simply disappearing. Sustaining a certain level of population will ensure the vital interests of disparate communities because the people are the lifeblood of the nation-state of Japan.

Although determining a precise required level would be impractical, a factor that should be considered is the balance of the population's age composition. If the aim is to avoid an aging rate of more than 40% and to stabilize it at 30% or below, then at least 80 million people would be roughly considered to be a "stable and appropriate population." Setting a floor at 80 million people is also not off the mark from the viewpoint of international status. In the case of Europe, the population of the major powers over the next few decades is projected to continue on the order of 80 million people—by 2050 the France will pass 70 million, making it the largest population in Europe alongside Germany—so 80 million people is a reasonable measure of major power status. In this regard, the Japanese government's Council on Economic and Fiscal Policy drew up a numerical goal in 2014 of "maintaining a population of 100 million in 50 years' time." This was the first time the Japanese government ever set a specific population goal at the political level. However, this aim is just a milestone on the road to stabilizing the population at 90 million people after 2090, and depends on achieving a population replacement level of 2.07 in 2040.

The reality, however, is that there is a strong possibility that the rapidly declining labor force will reduce the size of the economy, ushering in a long period of negative growth. Consequently, we need to be aware that, setting hope aside, maintaining a population of "100 million in 50 years" is quite difficult to achieve in practical terms. Indeed, even if the population did fall, it might be possible to maintain per capita national income if productivity per worker were raised. And increasing per capita productivity is an important policy issue in terms of an *adaptive* strategy for an era of population decline. In Japan's case, encouraging labor participation from women and the elderly is also vital. Setting aside the population issue, augmenting per capita GDP through innovation would maintain and improve Japan's vitality and appeal. Therefore, the adaptive strategy of maintaining per capita national income, meaning living standards, should be a key policy goal.

However, the size of the population will continue to be an important element for the state of the nation. Currently, Japan's population is the seventh largest among the G20 countries after China, India, the United States, Indonesia, Brazil, and Russia; its GDP is third after the United States and China; its per capita GDP is seventh after Australia, the United States, Canada, Germany, France, and Britain; and its exclusive economic zone is sixth in size after the United States, Australia, Indonesia, New Zealand, and Canada. All these meet the basic requirements of a major power.

If depopulation accelerates in the future and the scale of the economy is significantly reduced, it will likely be difficult to preserve Japan's current status as a major power even if per capita national income is maintained. Even if per capita national income was stable, the systems that guarantee public safety and peace of mind are based on national defense, public security and disaster response, all of which require a certain aggregate population. Additionally, if future generations, who will pay for past debts, are small, the per capita burden will increase, and so a certain population level—in particular, the working-age population—will need to be maintained to ensure public safety and living standards. Otherwise, both the population and per capita income will dwindle, leading to a strong risk of a descent into sudden poverty.

To overcome the various limiting factors that can occur in times of depopulation, both the government and the private sector will have to experiment with bold innovation. This is precisely where Japan can show its prowess as a problem frontrunner. It should be noted in creating population policy that aging should not be depicted as the enemy. Although there is a strong fear that as the population declines social costs such as pensions, medical care, and elder care will disproportionately grow, forcing a large financial burden upon each worker, and that this will have a negative impact (a population onus) on the incentive to work. However, aging statistics indicate that the average lifespan in Japan continues to increase. In other words, Japan is becoming healthier and more cultured. The labor participation of the elderly will become increasingly important in the two-pronged strategy of "mitigating" the population decline as described below, as well as "adapting" to it. There is a surprisingly large number of the elderly in Japan possessing superior abilities and who are in leadership positions. They are not an encumbrance. Despite being assets Japan should be proud of, more than half of those 60 years of age or older desiring reemployment are unable to find positions due to "age requirements."[6] How should their motivation be put to work for society? This question is nothing less than a national challenge in our era of population decline.

1.3.2 "Mitigation" of or "Adaptation" to Population Decline?

The "mitigation" faction warns that while it will be impossible to stem the population drop in the near future, Japan may be headed toward a catastrophic decline unless unremitting efforts for population recovery are pursued.

The "adaptation" faction proposes that, rather than wasting resources on realizing unachievable population recovery targets, policies should be firmly based on creating a new low-growth socio-economy that adapts to the long-term depopulation trend.

As in the preceding section, this debate also revolves around searching for an appropriate goal for population policy, but encompasses both goals and means. However, there is no need to view mitigation and adaption as diametrically opposed ideas. Both poles should be considered important and capable of producing greater policy results if pursued in tandem.

If the birth rate remains constant, the population will approach zero. Therefore, only by setting new targets to replenish the total fertility rate (TFR) will the first become possible. Mitigation is essential. We also need to bear in mind that depopulation varies significantly from region to region. In regions where population decline is already rapidly advancing, adaptation is the immediate issue and a new social system must be built to respond to this depletion. It is also possible to pursue adaptive strategies when safety measures are in place by creating a framework for relocating people away from landslide danger zones, where residential areas were developed by carving out mountainsides during the decades of population expansion. What is required is to create this kind of added value and linking it to mitigation.

One possibility for a simultaneous breakthrough combining mitigation and adaptation lies in technological and management innovation, but we should not entertain the illusion that innovation alone can solve the most human of problems in society. This belief would only interfere with a sincere commitment to the population issue and we need optimism braced against cold realism.

1.3.3 To Set a Birth Rate Target (Plus Population Target) or Not?

Those in favor argue that it is difficult to achieve results and verify the impact if policy goals are not concrete, therefore birth rate targets should be set. Any such target would merely be a goal to strive towards, not something enforced by the state or forced upon people.

Opponents argue that the decision to have a child is the most sacred right of the individual and stands at the core of human rights. It is not the purview of government. Therefore, the government should not set a desired fertility rate.

The greatest hurdle for Japan's population policy is the negative legacy resulting from the violation of this sacred right under the World War II-era "Beget and Multiply" policy. Even 70 years after the war, this moment remains a dark memory for the public, and for a long time the government was not free due to this history. According to the Ministry of Health and Welfare (the current MHLW) Establishment Law, population policy falls under its jurisdiction, but for many years the ministry did not have a "bureau in charge." After the war, the government of Japan may have used the word "raise" in regards to children in its policy objectives for childcare support, but it has carefully avoided the words "to bear." This wording is an example of a deep mistrust of government fertility programs, and that distrust has not dissipated. Any population policy is bound to fail if the underlying motive is to make women work or bear many babies because the population is falling. The government must pay close attention to avoid amplifying this lack of confidence.

Another factor behind this issue is the conflict in values regarding whether parents or society raise children. Under the Abe administration, the tax deduction for spouses is undergoing review in the belief that there can be no economic growth unless female employment is promoted. However, the Extraordinary Commission to Protect Family Ties in the ruling Liberal Democratic Party is strengthening its opposition to this in order to respect housewives who stay at home without a professional career but with the aim to protect and look after their family. In response to the Prime Minister's Office, which argues that they are only "changing the tax system so women who want to work can work as much as they like," the Commission counters that such a tax increase penalizes housewives. The origins of the dedicated housewife are partially found in the *samurai* social system. However, for ordinary people living outside of this system, there was no such thing as a dedicated housewife, and in the Edo period as well as in the present, common people have usually held an image of housewife as a multi-functional worker. In this way, the dedicated housewife could be thought as a social role derived from an archetypal of a certain class at a certain time (Oguchi et al. 2014).

The bottom line is that both parents and society raise children. However, public awareness of this has gradually been changing. In recent years, more than 92% of citizens agree that "society as a whole should bear the burden of having and raising children."[7] There was also another survey that suggested the ideal number of children for Japanese couples is 2.4, according to data from the National Institution of Population and Social Security Research.[8] This indicates that the Japanese population still has some latent capacity for recovery. It is true that the government cannot direct people to "have children" as a public policy, but it can, by contrast, reach out to people who want to "give birth." The government should play a greater role in realizing social and family systems in which couples can have their desired number of children. The core of such a system would be a work–family support system for both men and women to work professionally and raise children.

The mobilization of women for the labor force as an adaptation to population decline may well rob women of their psychological and practical leeway and lead to a further reduction in the birth rate. We need to focus on the significant impact a woman's childbearing age has on the low birth rate trend. In ultra-low 1.2 to 1.4 birth rate countries such as Japan, Germany, and South Korea, the main childbearing age for women is 30 to 34. On the other hand, in countries with a birth rate of 1.8 to 1.9 such as the Netherlands and Denmark, childbearing ages of 25 to 29 remain high, and in the United States with a birth rate of 2.0 the childbearing ages of 20 to 24 are also considerably high. In short, the earlier a woman gives birth, the higher the birth rate. These rates indicate that the life course of young people in their twenties and thirties has a significant impact on a nation's birth rate. In Japan, fewer people in their early to late twenties are opting to have and raise children.

Various public opinion surveys suggest that many young Japanese feel unable to tackle work enthusiastically and fear they will not be valued if they do not put in long working hours. More and more young people have little self-confidence and are unable to take the steps toward work or marriage.[9] The sense of fatigue engendered in Japanese workers—especially young workers with long commutes, extensive overtime, and ambiguous personnel evaluation systems—must be eradicated. A social system that provides the time to raise children as well as it promotes social advancement is required. Measures that assume women will do the child rearing need to be scrutinized from the ground up and switched to child-care measures that are gender-neutral. For example, the *Maternal and Child Health Handbook*, issued by local municipal offices to every mother providing them with free health checkups, counseling, and various informational services, should be renamed the *Parental and Child Health Handbook*.[10] Furthermore, we need to create a society where it is fun for the whole family to spend time with their children. The fact that Okinawa has Japan's highest birth rate and Fukui Prefecture with its tradition of "the whole family," both young and old, participating in shared work also has a high birth rate tells us we have forgotten something important.[11]

Being well aware of the difficulties involved in this issue, we would like to make the following recommendations.

After providing a clear indication of its strategy for creating new family and social systems, the government should set a TFR target (and population size) that references the ideal number of births (2.4 children per couple). It should also promise to take responsibility for creating a society where young people have the time and financial leeway to become pregnant, give birth, and raise children.

1.3.4 Are Overseas Workers a Benefit or a Cost?

To maintain Japan's per capita national income in an era of globalization, there is no choice but to supplement the employment of women and the elderly with skilled labor from abroad. Overseas human resources coming to Japan would further help mitigate the population decline.

However, opposition is deep-rooted and objections run as follows. Bringing in workers from abroad might be an economic plus for companies facing labor shortages, but it is the public at large that will bear the social costs, which is unfair. In addition, while workers from abroad may be a net positive in the short term, they will eventually prove costly in the long term by causing deteriorating security, religious and racial conflicts, as well as exacerbating a social divide.

Of all the mature democracies, Japan is unusual in that it effectively excludes immigrants. Immigration in Japan is almost a political taboo. However, immigrants and foreign workers should be considered as highly skilled professionals rather than short-term visitors, and a better environment for them to work and stay in Japan is of the utmost importance. This would require providing them support for living, vocational training, and education. In that sense, we stand in favor of introducing overseas human resources. However, injecting labor from abroad is not a magic wand for solving Japan's population issue and it would be difficult in practical terms to attract enough foreign workers to solve it.

While these human resources are indeed necessary for Japan to thrive in the world, the lifestyle of the future is likely to be a borderless one where "the individual chooses the country"—and many Japanese will also move abroad. Still, by opening its doors to foreign talent, the number of new compatriots studying, working, and building families in Japan will increase. In such a scenario, both the benefits and costs of an open-door policy should be calmly weighed to contribute to the mitigation of and adaptation to an aging and declining population. Such a policy can be maintained as long as the benefits exceed the costs.

However, we need to take into account the question of who benefits and who pays the costs as well as the changing dynamics over the short, medium, and long term. The major beneficiaries will be companies plagued with labor shortages, especially in industries where structural reform and overseas relocation is difficult, such as the elderly care business and their clients and families. Should these companies turn a profit, in the short term at least, the benefits will outweigh the costs. However, social costs such as education and social security will increase in the medium to long term especially if such workers bring their families to Japan. Here again the issue of overseas labor can be placed under the scope of both the mitigation and adaptation strategies. With a declining population, those companies unable to reform structurally or relocate overseas will be culled. In the short term, the demand for unskilled workers for these industries will also diminish.

In the future, overseas workers will not consist of low skilled but rather highly skilled workers and professional specialists who can contribute to the innovation essential for Japan to survive its depopulation phase. Above all, our efforts should focus on highly skilled workers. In the case of outstanding researchers, engineers, artists, and others, the benefits of a long-term presence, including the external impact on their surroundings, are great even if such individuals do not become permanent residents. As a framework leading to future settlement in Japan, the development of employment and career path systems for exchange students in Japan should be put in place. The current number of foreign students in Japan is just under 170,000 (2013). While the share of those hoping to work in Japan after graduation is 65% of privately financed students, only about 40% actually find employment.

At the same time, the requirements for foreigners to acquire permanent residence should be relaxed. The highly skilled are already the object of international demand. Universities, which act as their incubators, are also subject to intense international competition. Success will depend on how far Japanese society and higher education can open their doors to the world, how far Japan can improve its competitiveness; in other words, how capably Japan can accept high-quality labor from the overseas. At the same time, regional economic integration is required to relocate human resources in the Asia-Pacific region, as well as widening the range of domicile selection. Economic integration is capable of mitigating the impact on a society heading toward an aging and declining population (Menon and Melendez-Nakamura 2009). Preparing the social infrastructure and social psychological groundwork is essential for accepting overseas labor not simply as workers, but as fellow residents. A strong-rooted fear of immigration in certain parts of Japanese society forgets the fact that Japan proactively accepted highly skilled workers from different ethnic groups in ancient and early modern times. Many of those coming to Japan settled here thanks to a welcoming system of acceptance and this settlement process is believed to have been completed by the Nara period (710–794). Records also show that as of the ninth century, about a third of the clans living in Kyoto and Kinai, around the Kyoto area, were from abroad.

1.3.5 Tokyo or the Regions?

A large number of women aged 20 to 39 leave the regional periphery to work in Tokyo. Some people see this as contributing further to regional extinction. They preach a mantra that the drop in regional population needs to be halted by rectifying the over-concentration in Tokyo, and redistributing wealth, income, and employment to the regions.

On the other hand, there are people who think the further development of Tokyo as a global city is required. They counter that young people amass in Tokyo because the regions have failed to provide attractive places of employment and that preaching decentralization from Tokyo is futile.

In 2014, the population influx to Metropolitan Tokyo from other areas increased from the previous year, with incoming residents exceeding outgoing residents by more than 110,000 in the space of just 1 year. Looking at changes in the working population by industry in recent years, the number of workers in tertiary industries such as information and communications, academic research and professional services have increased in the three major metropolitan areas of Tokyo, Osaka, and Nagoya. In comparison, the only field increasing in the outlying regions has been medical care and social welfare. The hollowing out of industry in rural areas is rapidly gaining ground.

Further contributing to the lack of employment in the regions is the concentration of most large-company headquarters in Tokyo. In particular, over-centralization in Tokyo vacuums up the young from the regions, accelerating “regional attrition.” Should we foster a U-turn away from Tokyo? Or should we consolidate shrinking local governments, promote mergers between core regional cities, and expand urban city areas? From the point of view of disaster countermeasures, starting with earthquakes, Tokyo’s over-centralization is undesirable. Serious consideration should be given to a temporary transfer of the capital’s functions to the regions. Also, even Greater Tokyo had an aging rate of 24.2% in 2015, and the central Tokyo wards will face a population decline in 2020. This downturn will alleviate overcrowding problems such as housing shortages and traffic congestion that are the fate of megacities.

There is a possibility that new problems will occur. Although apartments built in Metropolitan Tokyo during the high-growth period are rapidly getting older, rebuilding has been negligible. The slumification of apartments is likely to become a significant problem. Moreover, medical and care services in central parts of Metropolitan Tokyo have fallen into a serious shortage.

But the phenomenon of Tokyo’s over-centralization should not be seen as a battle of Tokyo versus the regions. The rapid aging and depopulation taking place today in the regions will one day occur in Metropolitan and Greater Tokyo. We should take the view that both the regions and Tokyo face the same challenges. A weakening Tokyo does not correspond to a strengthening of the regions. The only way to strengthen the regions is to create strong cities, strong businesses, strong research institutions, and strong societies in regional locations. Only then will it be possible to pursue a multipolar vision that promotes multiple global (or Asia-Pacific) cities with Tokyo at the apex. Tokyo is one of the world’s leading global cities and its competitive advantage and appeal needs to be further enhanced. This reconsideration would also help attract overseas human resources to Japan.

1.3.6 Intergenerational Benefits and Burdens

The intergenerational benefits and burdens of social security have become inequitable due to the aging and declining population. In addition, the elderly majority has turned into the most powerful political lobby, and with the rise of a "silver democracy" that accords its own class preferential treatment, there is a strong possibility that intergenerational injustice will become more structurally embedded.

Are the elderly going to exploit the young in a tyranny of the majority? Is that something to be overthrown by the young? Or is dialogue and cooperation to rebuild intergenerational equity possible?

A Japanese baby that has just uttered its first cry is from the moment of birth is already carrying a debt of 8.11 million yen. While the elderly will receive social security benefits in the form of pensions, medical, and elder care, the young and those not even born bear the burden. The principle equity of benefits and burden between the generations is currently under threat. With an aging demographic, all of the mature democracies that have created welfare states, Japan included, are entering the era of a silver democracy. In these countries, for two centuries the young have been the main beneficiaries, but now they are poised to be the main losers in the coming era.[12] All existing political parties, whether they are on the right, left, or in the center, are competing with each other to promote the protection of elderly benefits, young people feel a sense of alienation from party politics, raising the danger of them being captured by far-right parties.

Japan is not free of such risks. Consequently, in the face of the overwhelming political power of the elderly, we need to address the question of how to divide the pie with the younger and future generations. And this pie needs to be redistributed even as it shrinks. This is unlikely to be achieved by a revolution or riots by the young to overthrow the tyranny of the elderly. We think the answer is sharing the challenge through an intergenerational dialogue rather than intergenerational conflict. We should recall the insight of the political thinker Edmund Burke, who pointed out in his *Reflections on the Revolution in France* that the true social contract is a partnership between the generations to preserve sovereignty across time (Burke 1982 [1790]).

Such altruistic spirit and motivation exist in human society, and as a whole society should devise ways of resolving these issues. And when doing so, the family and community should be the most important unit of intergenerational dialogue. The family cannot exist in isolation from society, so therefore neither can intergenerational dialogue within the family fully occur separately from society. The family is also an institution to be supported by society—especially local communities. What is needed is a new social perspective that emphasizes intergenerational cooperation and new family values to carry out work and childcare in which men and women equally play a role. The key resource for a new set of population policies is to shift the entire social system to one that prioritizes pregnancy, childbirth, and childrearing. No solution is possible unless the challenges from all the negative chain reactions are first met. In that sense, creating a population policy should be considered as a new form of nation building. It is a matter of course that politics and government must tackle establishing this with unflagging resolve.

Ultimately, however, the population issue belongs in the domain of the individual and family hopes and intentions. It cannot be left up to politics and the government, nor does it lend itself well to politics and government. These are issues where by rights we should not harbor excessive expectations for politics and government. It is an issue that will fail to bear fruit without the understanding, empathy, and commitment of the entire nation. What should be required is not reform from above, but a national awakening.

1.4 Book Layout

Chapters 2, 3, and 4 serve as an overview of the current situation of population decline in Japan.

Depopulation is often deemed to be a regional problem, but the phenomenon is already affecting Metropolitan Tokyo. Chapter 2 concludes that the change is more serious in urban areas, where the sense of local community is weaker, than in regional areas when the problems of an aging and declining population are advancing simultaneously. The signs are already apparent with the elderly meeting lonely deaths, an increase in rundown (marginal) apartments and vacant houses, and a shortage of elder care facilities. The situation is also becoming increasingly serious in suburban Tokyo, where new town developments were promoted to provide the dream of home ownership in the high-growth period, in what may be termed the "Metropolitan Shock."

Chapter 3 illustrates that the situation is already serious in regional areas. These areas, which have experienced three population outflows since the high-growth period (1962–1972), are not only experiencing changes in population structure due to a drop in birth rates and aging, but face industrial decline, shrinking economies, and population loss all combining to make life more difficult for residents. This regional decline has led to a withdrawal of such daily life infrastructure as supermarkets, shops, and gas stations. The impact is also being felt in schools, elder care institutions, and nursing services, ushering in an era of regional reduction, or "community collapse."

Chapter 4 explains how the impact of population decline has emerged as an economic constraint on both the production and consumption sides of the macro economy. On the production side, the drop in the labor force means labor productivity needs to be raised in order to maintain growth. Realizing this will not be easy, judging by past growth rates. Given that as people age they tend to liquidate their household savings, an aging population will have a negative impact on the economy by reducing investment. In addition, as an aging and falling population continues apace, the impact on consumption will be severe. There is a possibility of fiscal crisis due to higher tax and social security burdens on the one hand, and a conceivable drop in consumption due to the working generation's growing anxiety about its burden and future on the other. If total factor productivity (TFP, an indication of the capital stock per worker and the degree of technological progress) can be raised, this would mitigate the effects of a shrinking labor force. However, the fact that labor productivity is low in the service sector, which accounts for a high industrial share, will make such mitigation measures difficult.

Chapters 5, 6, and 7 comprise the core chapters of statistical research because verification of the past is essential to glean lessons that contribute to the future.

In Chapter 5, we look back on population policy, noting that the current crisis is the result of a conscious avoidance of demographic policy even with the existence of countermeasures for a declining birth rate. Due to reticence in light of the history of the World War II-era "Beget and Multiply" program, the state has avoided policies promoting birth. Moreover, as aging increased as the population grew after the war, policy awareness was directed at aging rather than depopulation. As a result, policies centered not on the declining birth rate but on measures for the elderly. The fact that countermeasures for the declining birth rate began to be implemented in earnest after the collapse of the bubble economy when the public purse was tight was also harmful. This policy delay can be witnessed in relation to the foreign workers issue as well.

Chapter 6 focuses on national land policy in the regions as well as by the government. "Balanced national development" is national land policy, factory relocation and a system of regional public finance all aimed at universally spreading infrastructure investment and public finances to the regions as a response to population concentration in cities triggered by high growth. On the other hand, homes were provided by suburban housing estates and a policy of home ownership in urban areas where the population was expanding. This occurred under the stable government structure known as the "1955 regime" due to the leadership of the Liberal Democratic Party that began in 1955 and lasted until 1993. If we look back from the perspective of population policy, national land policy to date has corrected distortions caused by imbalances in the population structure. However, policies suited to an era of population growth need to be reviewed today with the shift to depopulation. The passage of time has exposed the weaknesses of urban planning like city sprawl, the harmful effects of divided ownership systems that impede the rebuilding of rundown apartments, and inadequate disaster preparedness and emergency medical systems. The effectiveness of existing policy measures is being questioned, including those for land planning, local government mergers, and using municipal bonds to attract businesses.

These above problems are closely related to the nature of democracy, which we analyze in Chapter 5. The issue of "silver democracy" is particularly salient in Japan, where the aging and population decline are advancing in tandem. If democracy is only manipulated by the will of the majority, it merely becomes a tyranny. To date, democracy has suppressed the tyranny of the majority through mobocracy by superimposing various measures on a foundation of representative democracy. Humans are selfish and rational as well as altruistic and irrational at the same time. Nevertheless, the population issue has continually been put off. In part, two causes behind the delayed response to the depopulation question are first, the absence of both fierce opponents and proponents, and second, the failure of policy entrepreneurs who would have led the push for changes necessary to switch policy amidst the silo mentality of government administrators. The greatest wisdom of democracy will be required to address the population question before it is resolved by a silver democracy intergenerational conflict.

A "vision and policy objectives" for resolving a society in population decline is a question posed in Chapter 8, 9, and 10. Each chapter addresses one of the three pillars; fiscal and macro-economic policies; optimal reallocation of population and land, area management and crisis management; and business innovation.

Demographic trends and fiscal issues are discussed in Chapter 8, noting that the key here is that Japan's public debt is soaring vis-à-vis social security costs and that social security is the greatest risk factor for public finances. The options for solving this problem are the troika of economic growth, tax increases, and spending cuts. It will be difficult to rebuild the public purse only through economic growth, and any attempt to stabilize public finances without suppressing social security will require substantial tax increases. Specifically, this involves an incremental increase in the consumption tax as well as a certain curtailment of social security benefits. But in terms of investment in the future, measures to counter the falling birth rate need to double support for childcare. Deploying foreign workers will also have to be considered using approaches in other countries as a point of reference.

Chapter 9 deals with the second pillar, the optimal reallocation of population/land, area management, and crisis management. City density must be increased along the lines of a compact city to maintain services in regional cities, although this will take time. Citizens have started to become conscious of the need to manage, maintain, and consolidate urban facilities and infrastructure after the Great East Japan Earthquake on March 11, 2011 (BBC 2012) and the Sasago tunnel accident on December 2, 2012. In light of this trend of decaying infrastructure, land use and density control are way behind in terms of developing urban facilities, according to the Ministry of Land, Infrastructure, Transport and Tourism (MLIT 2013), and it will be necessary in the future as the public demands the means for achieving this. To maintain regional living areas there must be fundamental changes to regional fiscal systems—including the system of local self-government and regional tax grants—and a course of vigorous regional cooperation must be pursued. Insufficient mergers and regional cooperation policy that occurred in the past needs to be reviewed, and regional systems need to be restructured to tackle regional tax grants, local government assemblies, and regional administrations. It should also not be forgotten that over-centralization in Tokyo also encompasses issues of crisis management for disasters and so on. Customization will be needed to maintain the reservoir function regional city areas play for large urban areas.

The third pillar, business innovation, is discussed in Chapter 10. A major cause of depopulation is that young people flow out to the big cities because they fail to find attractive work in the regions where they live. The most effective way of halting this trend is to create “stable jobs that are both challenging and provide appropriate income” in local communities. Chapter 10 sees this as a “chance” for the policy-issue frontrunner Japan to “demonstrate to the world” how to respond to the depopulation crisis.

Notes

1. IPSS 2012.
2. Economic Planning Agency 1992.
3. Cabinet Office 2015.
4. Council on Economic Fiscal Policy 2014.
5. OECD 2014.
6. Cabinet Office 2006.
7. According to the Cabinet Office’s survey on Japan’s future vision for population and socio-economy (Cabinet Office 2014a), 92.3% of respondents were in favor or slightly in favor of the idea that “the burden of bearing and raising children should be supported by society as a whole.”
8. IPSS 2011.
9. Cabinet Office 2014b.
10. MHLW n.d.
11. The TFR for Okinawa Prefecture is 1.94 and Fukui Prefecture 1.60 (MHLW 2014).
12. See Mazower 2014.

References

BBC. (2012, March 11). Japan quake: Loss and recovery in numbers. *BBC News*. http://www.bbc.com/news/world-asia-17219008, accessed on September 25, 2016.

Burke, Edmund. (1982 [1790]). *Reflections on the Revolution in France*. London: Penguin Classics, Reprint edition.

Cabinet Office, Government of Japan. (2006). White paper on the national lifestyle (Kokuminseikatsu hakusho 2006-nenban). http://www5.cao.go.jp/seikatsu/whitepaper/h18/10_pdf/01_honpen/, accessed on February 1, 2017 (in Japanese).

Cabinet Office, Government of Japan. (2014a). A public opinion survey on the future vision for Japan’s population and socio-economy (Jinko, keizaisyakaitou no Nippon no miraizou ni kansuru yoron chosa). http://survey.gov-online.

go.jp/h26/h26-shourai/index.html, accessed on October 5, 2016 (in Japanese).

Cabinet Office, Government of Japan. (2014b). 2014 White paper on children and youth (Heisei 26nen Kodomo Wakamono Hakusho). http://www8.cao.go.jp/youth/whitepaper/h26honpen/, accessed on September 25, 2016 (in Japanese).

Cabinet Office, Government of Japan. (2015). Annual Report on the Aging Society: 2015. http://www8.cao.go.jp/kourei/english/annualreport/2015/2015pdf_e.html, accessed on September 25, 2016.

Council on Economic Fiscal Policy, Expert Panel. (2014). Report from Growth and Development Working Group: Sustaining population, Innovation and Branding Japan with avoiding reducing spiral (Seicho · Hatten wa-king gru-pu houkokusyo: syukusyo supairaru wo kaihi shi, jinko anteika · inobeisyon · nippon buranding) http://www5.cao.go.jp/keizai-shimon/kaigi/special/future/1114/shiryou_01_2.pdf, accessed on February 14, 2017 (in Japanese).

Economic Planning Agency. (1992). White paper on the national lifestyle (Kokumin seikatsu hakusho). http://www5.cao.go.jp/seikatsu/whitepaper/h4/wp-pl92-01503.html, accessed on October 2, 2016. (in Japanese).

IPSS. (2011, October 21). An outline of the results of the nationwide survey and couples survey on marriage and childbirth. 14th Basic survey on birth trends, 2010 (Dai14kai shusseidoko kihon chosa). http://www.mhlw.go.jp/stf/shingi/2r9852000001thzi-att/2r9852000001ti0z.pdf, accessed on October 4, 2016 (in Japanese).

IPSS. (2012). Japan's estimated future population (estimated in January 2012, medium variant projections) (Nihon no shoraisuikei jinko, 2012nen 1gatsu suikei), http://www.ipss.go.jp/syoushika/tohkei/newest04/sh2401smm.html, accessed on February 21, 2017 (in Japanese).

Mazower, Mark. (2014, February 6). Baby boomers have blighted their children's prospects. *Financial Times*. http://www.ft.com/cms/s/0/3af689d4-81cd-11e3-87d5-00144feab7de.html, accessed on September 25, 2016.

Menon, Jayant and Anna Melendez-Nakamura. (2009). Aging in Asia: Trends, impacts and responses. Asian Development Bank Working Paper Series on Regional Economic Integration No. 25. https://www.adb.org/publications/aging-asia-trends-impacts-and-responses, accessed on September 25, 2016.

MHLW. (2014). Vital Statistics Summary 2013 (2013-nen Jinkou Doutai Toukei kakutei chi no gaikyou, Hyou 3-2). http://www.mhlw.go.jp/toukei/saikin/hw/jinkou/kakutei13/dl/07_h3-2.pdf, accessed on August 7, 2017 (in Japanese).

MHLW. (n.d.). *For Your Well-being during Pregnancy and Delivery* (information sheet). http://www.mhlw.go.jp/bunya/kodomo/boshi-hoken10/dl/02.pdf, accessed on September 25, 2016.

MLIT. (2013, June 18). Report on the investigation and meeting on the Sasago Tunnel accident (Tonneru tenjoban no rakkajiko ni kansuru chosa kentoiinkai). http://www.mlit.go.jp/road/ir/ir-council/tunnel/pdf/130618_houkoku.pdf, accessed on September 25, 2016 (in Japanese).

OECD. (2014, December). Focus on inequality and growth. https://www.oecd.org/social/Focus-Inequality-and-Growth-2014.pdf, accessed on February 2, 2017.

Oguchi, Yujiro, Ryuichi Narita & Sanae Fukuto (eds). (2014). *A new system of Japanese history 9: The history of gender (Shintaikeinihonshi9 jendashi)*. Tokyo: Yamakawa Publishing.

Piketty, Thomas. (2014). *Capital in the twenty-first century*. Cambridge, MA: Belknap Press of Harvard University Press.

CHAPTER 2

The Greater Tokyo Shock

Seta Fumihiko, Otake Hiroshi, and Umeyama Goro

Japan has become a society with an unambiguously declining population, currently at 127.11 million. According to the 2015 Population Census published by the Ministry of Internal Affairs and Communications (MIC), the Japanese population dropped for the first time since the Census was first taken in 1920. Lifestyle-related industries (retail, food and drink, entertainment, healthcare institutions, etc.) are expected to shrink as a result. For example, only municipalities with a population of 27,500 or more have an 80% or higher chance of housing a general hospital (for those with 5500 or more, a 50% or higher chance), and some of these may lose their hospitals completely. The service sector, which accounts for 60% of regional employment, could also weaken. The proportion of vacant homes has consistently grown, from 9.8% of all homes in 1993 to 13.5% in 2013.[1]

Seta. F (✉)
Department of Urban Engineering, Graduate School of Engineering, University of Tokyo, Tokyo, Japan

Otake. H
Department of Anesthesiology and Critical Care Medicine, Showa University School of Medicine, Tokyo, Japan

Umeyama. G
Consultant, Risk and Crisis Management, Tokyo, Japan

Y. Funabashi (ed.), *Japan's Population Implosion*,
https://doi.org/10.1007/978-981-10-4983-5_2

Lifeless towns are gradually emerging—even in metropolitan Tokyo. While the earlier description of the housing and social situation might not be occurring everywhere at the same time, a very bleak scenario awaits the nation if no effective measures are taken and current demographic trends continue. The problems that a declining population and the ensuing community decay and deterioration in the living environment could trigger tend to be viewed in insolation as a regional problem. However, as an ongoing event and a factor that will have an increasing influence on Japan's future prospects as a nation, depopulation is also stripping towns in the Tokyo suburbs of their energy to thrive.

This prognosis is not fiction; it is a reality that is currently unfolding right before our eyes.

2.1 "We'll Be a Marginal Estate in Ten Years' Time"

> "They say, 'home is where you make it', but I'm not sure if I can keep living here."[2]

Hatoyama New Town in Hatoyama, Central Saitama Prefecture, is about an hour away by rail and bus from a major hub in northwest Tokyo, Ikebukuro Station. Suzuki Taro (a pseudonym), a man in his mid-seventies who has lived in this residential town developed in the high-growth period of the 1960s for more than 40 years, despairs for the future.[3] Hatoyama New Town consists of some 3200 houses.[4] Although a downturn due to age is indeed noticeable in the buildings themselves, each plot is generally large and the town blocks are well appointed. Many of the residents worked for large companies and the central government in Tokyo, so the percentage of high-income earners is said to be higher compared to other contemporaneously developed suburban housing estates, such as Tama New Town, Tokyo (Bureau of Urban Development, Tokyo Metropolitan Government n.d.). Suzuki worked until retirement age in leading companies in downtown Tokyo.

What concerns residents is the increase in vacant homes and aging occupants.[5] On the block where Suzuki lives, residents keep moving out, creating a situation where 4 out of 16 houses now stand vacant. This vacancy rate does not differ much from block to block, and in some areas the number of residents has fallen by more than 20% in the decade since 2000.[6] Many of the residents have lived here since the establishment of the residential development, and nearly more than half are 70 years or older.

At the same time, the number of children continues to fall. One of the two primary schools in the area has already been shuttered. The remaining school building is now being used as a *crèche* (day care center) and rental classroom space, but the only people seen inside are the elderly engaged in hobbies such as flower arrangement (*ikebana*) and patchwork sewing.

This overall loss of residents and the aging of those who remain are two forces that are gradually making local life much more tenuous. A grocery store barely survives in the New Town center, but the large-scale retailer withdrew in 2014. The floors of the commercial facility built at the time of the estate development were initially crowded with a variety of shops, but are mostly vacant now. The private company, Japan urban development corporation, which developed this residential estate is already bankrupt, and there is very little momentum emerging from the "bed-town residents" themselves for regenerating their community.[7] Amidst tight town finances, a variety of support is provided from operating an on-demand taxi service to help with various yearly festivals, but even these efforts have not stopped the inexorable decline.

Weakening ties between residents have also further exacerbated the situation. In February 2014, the heavy snowfall in the region still blanketed the streets of Hatoyama New Town the next day and even the day after. Going out in this town of hilly inclines when the roads freeze in such weather is impossible. The municipality strove to remove snow from the main roads, but they apparently did not have the manpower to get around to the smaller streets within the residential development. Suzuki stated that he not only shoveled snow from the road in front of his own house, but also asked neighbors to clear away the snow that blocked their entrances. While some residents came to their doors, they made a variety of excuses and in the end did nothing. Estates where residential associations can no longer function and the necessary staff and facilities to sustain a harmonious living environment cannot be maintained due to aging are referred to as "marginal estates." While the term "marginal community," referring to increased aging and isolation in mountainous areas, is firmly established in the lexicon, "marginal estates" refers to the urban version of the same nationwide phenomenon and is linked directly to a declining population.

2.2 The Shrinking Metropolis

Population in the Tokyo metropolitan area will also fall in the future. Although the timing of population decline will start later than rural areas, the near future is already clear. According to population estimates by the

National Institute of Population and Social Security Research (IPSS), in 2015, the population will peak even in Metropolitan Tokyo (Tokyo, Saitama, Chiba, and Kanagawa Prefectures), the central sectors of what is called Greater Tokyo, then fall steadily from the year 2020.[8] Looking at census figures for fluctuations in prefectural population, the steepest decline between 2005 and 2010 was in Akita Prefecture (−5.2%) followed by Aomori Prefecture (−4.4%). These examples show the harsh reality northeastern Japanese prefectures face with increasing threats of depopulation and the worry of what do with an increasing number of marginal communities.[9]

In fact, even in the metropolitan area there are places at the municipal level where the population has fallen as fast or faster than the northeastern region. Choshi in Chiba Prefecture, for example, where the closure of the City Hospital created an uproar in 2008, is experiencing depopulation at a much faster pace than Akita or Aomori Prefectures.[10] Although the hospital itself managed to somehow resume operations in 2010, the outlook is bleak. Elsewhere, in Yokosuka, located in Kanagawa Prefecture at the tip of the Miura Peninsula, a population outflow continues making it the place with the greatest population decline in 2013. Although the rate of decline was less than 0.5% year-on-year, the city finds itself in a situation where depopulation continues unabated and some areas are already becoming marginalized communities. In this way, metropolitan Tokyo is also progressively being eroded from its periphery. Turning one's gaze even to commuter areas that feed Tokyo, the population in Satte, Saitama Prefecture, which is located on the metropolitan edge, is dropping at around the same rate as Akita and Aomori Prefectures. Even in other commuter suburban municipalities, such as Kasukabe in Saitama Prefecture, Abiko in Chiba Prefecture, and Zushi in Kanagawa Prefecture, predictions put the pace of population decline in one or two decades on a par with Aomori Prefecture. The same situation is predicted for places such as Tokyo's Adachi Ward and Chiba City (Fig. 2.1).

The aging of metropolitan inhabitants is now even more serious than the problem of population decline. The prefectures with the highest aged population ratio are Akita (29.6%), followed by Shimane (29.1%). The situation in both prefectures, where between one-third and one-quarter of people are 65 years of age or older may seem to be an issue completely unrelated when viewed from the youthful image of metropolitan Tokyo. However, Chiba Prefecture, next door to Tokyo, is predicted to reach the same level around 2022, and Saitama Prefecture around 2028 (Fig. 2.2).

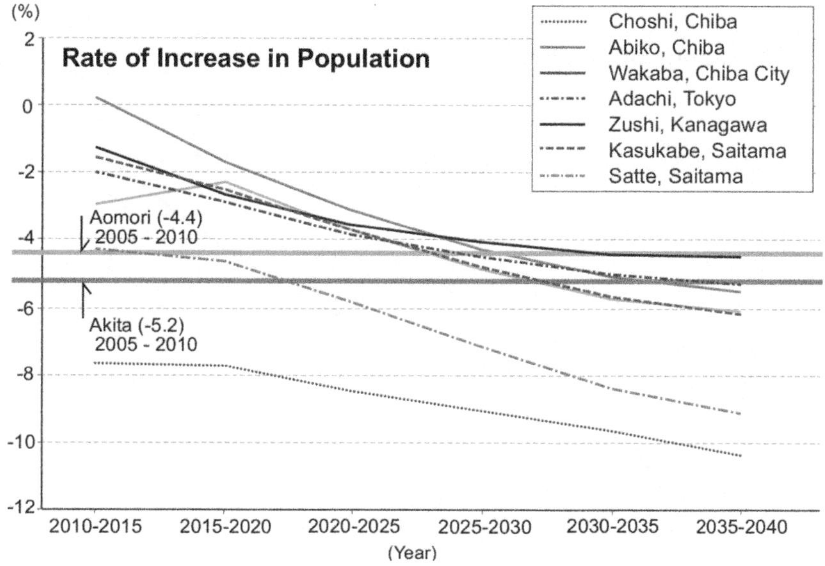

Fig. 2.1 Population forecasts for local municipalities in Greater Tokyo and current population trends in the regional area. Source: MIC, National Census 2010; IPSS 2013

Even in the metropolitan area, a similar degree of aging in regions with marginal communities is fast approaching. An aging population also means a drop in the number of young people who will support the regions in the future. In fact, local Tokyo suburban municipalities already find themselves in dire straits. Looking at the aging rates for 2010, many suburban municipalities are currently above the national average (23.0%), such as Saitama Prefecture's Hatoyama-cho (28.2%), Kanagawa Prefecture's Kamakura (27.6%), and Chiba City's Wakaba-ku (25.6%). Since young people in their teens and twenties flow from the outlying regions into the metropolitan areas for education and employment, an increase of this population cohort in local governments in the vicinity of Tokyo should be large. Moreover, because they are also suitable for residential development, an increase in this demographic can also be expected as they marry and buy homes.

By contrast, however, the reality is that the population in their teens and twenties continues to fall in Tokyo's suburban municipalities. In the period 2005–2010, only 13 of the municipalities in Greater Tokyo

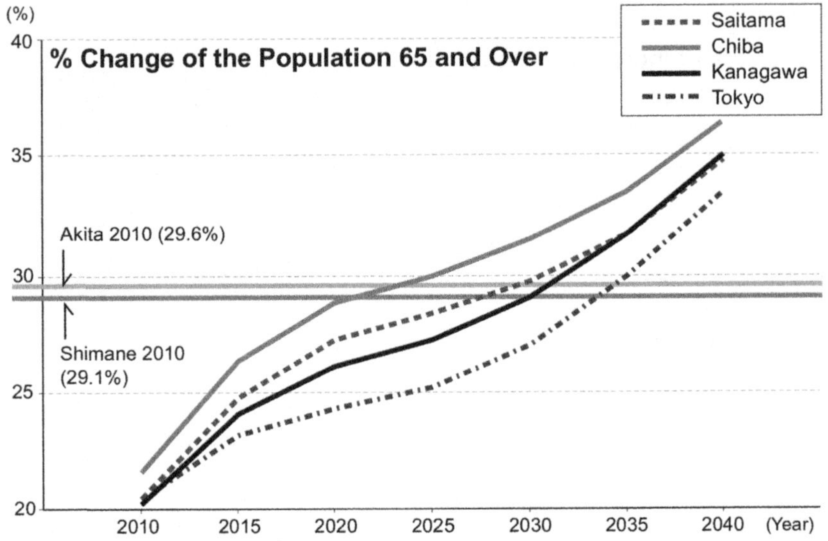

Fig. 2.2 Aging forecasts for Greater Tokyo and current aging in the regional area. Source: MIC, National Census 2010; IPSS 2013

excluding the outlying islands—such as Toda in Saitama Prefecture and Mitaka in Tokyo—experienced an increase in this population stratum.

2.3 Vacant Houses, Marginal Apartments, Consumer Refugees

The Tokyo metropolitan area will shrink as an aging and decreasing population increases, and there is great concern that suburban municipalities no longer encompassed by Tokyo can deal with the numerous problems that will emerge along the way, as detailed below.

2.3.1 Vacant Homes on the Rise

Even in areas where the population is falling, to some extent demand for new homes will continue because children grow up and want their own homes and employees move and relocate. A waste of building resources and land can be avoided if demand for new homes is met by using existing housing stock and new construction in existing urban areas. However, the actual case is that currently new, cheaply priced and loosely regulated land is being developed. As such, housing still in a good enough state and vacant units in well-developed housing estates go completely ignored and underutilized. Even with a population decline, unregulated development continues to advance, resulting in an over-supply of housing stock. Consequently, the number of vacant houses in existing urban areas climbs higher.[11] An increase in vacant homes becomes a larger social and political problem in terms of security and disaster prevention. Lower density in local areas also leads to issues such as rising infrastructure maintenance costs and the emergence of so-called consumer refugees (***kaimono nanmin***), or people living in what are called "food deserts" (Kidachi 2011). Local communities will be devastated if the number of inhabitants who have no interest in community life grows as people who own land and houses but are unable to live in them rent these properties out or merely leave them vacant.

The number of vacant houses with unclear ownership is also on the rise. If owners' rights remain as strongly protected as they currently are, neither the police nor local authorities can do anything about empty houses that fall into a dangerous state of disrepair. Moreover, the security of the community and problems of squatters moving in illegally are also potential future conundrums. According to a report by Yoneyama Hidetaka of the Fujitsu Research Institute, vacancy rates will continue to rise generally across the nation in the future (Yoneyama 2012). It will be difficult to eliminate these vacancies unless the excessive housing stock is significantly reduced and the use of existing housing is promoted. If the number of new buildings and old houses disappearing through a lack of repair stay at their present levels, the vacancy rate in 2028 will reach 23.7%, or almost one house in four. This represents a rise of slightly more than 10.6 points from the 13.1% rate recorded in 2008. As of that year, Yamanashi Prefecture had the highest vacancy rate of 20.3%. Although there were some municipalities on the Tokyo outskirts with high vacancy rates, such as Zushi in Kanagawa Prefecture (17.8%), Kisarazu in Chiba Prefecture (15.3%), and Fussa in Tokyo (14.4%), the future will see many more municipalities dealing with far higher vacancy rates.

2.3.2 "Marginal Apartments" Emerge

A major concern for Japan is that population decline will have a far more serious impact on apartment dwellers than those who own their own

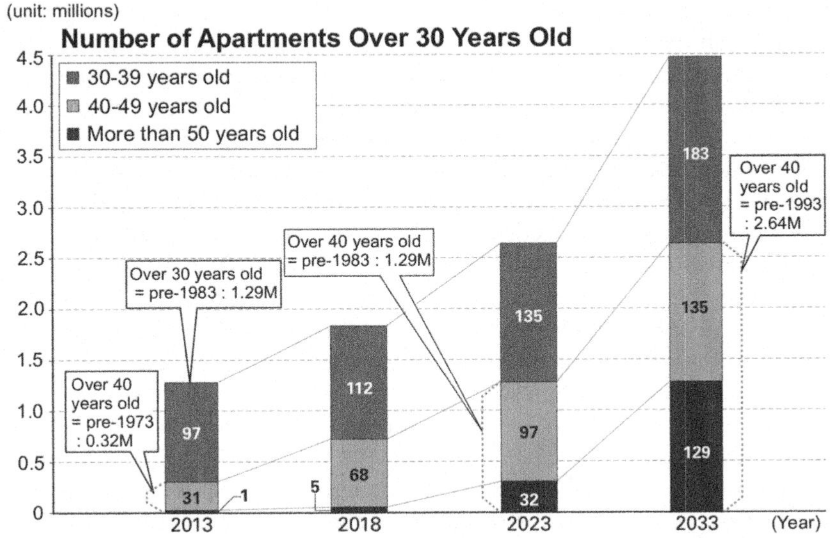

Fig. 2.3 Projected increase in old apartments. Source: MLIT, Facilitation of the Reconstruction of Apartment Buildings 2013

detached homes. Many of the apartments in the metropolitan area were built 30 to 40 years ago and will shortly enter a phase where they have to be renovated or rebuilt, as is fairly common in Japan. These days, when the population is in decline and houses are in abundance, new tenants are unlikely to be found even if apartment buildings are rebuilt. There will not be that many residents willing to rebuild given the considerable costs. Coupled with the fact that rebuilds require the agreement of more than 80% of all building residents, marginal apartments left to rot will crop up with a greater frequency, even if the time for rebuilding has arrived (Fig. 2.3). This issue will be revisited in detail in Chapter 6.

2.3.3 *No More Schools*

Schools are the "faces" of local communities, often serving as a central gathering spot for both young and old. Buildings and facilities are now consolidated as student numbers drop, so schools as community outlets will also decline steadily in the future. In Japan, elementary and junior high schools also serve to strengthen local community ties by hosting social events such as sports days and school plays. If schools close, communities will lose a force that offers them a keen sense of unity—they will lack a lynchpin for social cohesion. Currently, it is not uncommon for obsolete school buildings to be left intact and to be reutilized as new public facilities, such as community activity offices and nursing homes. When the location is good there are cases where such facilities are sold to developers as a vacant lot and redeveloped into apartment blocks. Sales to the private sector, however, raise strong concerns among residents that the heart of their community will be lost, leading to opposition movements throughout the country.

2.3.4 Public Transportation More Inconvenient

With the advent of the car society, and as train and bus usage falls, many public transportation operations will no longer be viable when the number of commuters drops according to data from the Ministry of Land, Infrastructure, Transport and Tourism.[12] A decrease in passengers will reduce service levels, such as the number of lines and operating frequencies, creating a vicious cycle that further reduces the number of passengers. To date, this problem was thought to be the exclusive domain of rural areas and provincial cities that were slowly emptying out. However, the fact is, these issues are already starting to become a reality even in metropolitan areas. In 2013, the Japanese media reported that Seibu Railway, a company where foreign investment firms hold lots of stock, was considering a strategy of abolishing certain suburban branch lines.[13] This was a significant shock given that it was not only lines in mountainous areas like the Chichibu Line that were targeted, but also included lines running through urban districts on the outskirts of Tokyo which are well known as residential areas such as Musashino, Koganei, Kokubunji, and Tokorozawa. The investment firms subsequently denied the report, but residents along the branch lines remained unconvinced and came out in opposition to the lines' potential abolition.

On the other hand, although many passengers using suburban lines linking the city center and the airport ride it from terminus to terminus, there are few passengers using the stations in between. Since competition between rail companies is fierce on airport routes, the railway company that runs this route ensures a fast train runs from the city center to the airport in its attempt to capture passengers by reducing the number of station stops en route at the expense of passengers who use the railway for everyday commutes. To deal with the inconvenience, residents along the line organized a vanpool by themselves that runs to the terminal station. In response to this kind of problem, local municipalities already provide support to railway and bus companies for existing lines as well as to operate community buses, on-demand buses, and demand-type vanpools. However, the drop in commuter numbers due to aging and population decline is an obstacle and it is hard to say that countermeasures have had a sufficient impact.

2.3.5 "Consumer Refugees" Emerge

When the number of residents falls, large shops such as supermarkets that are no longer profitable withdraw from the area. Small shops like individual retailers also follow suit. In addition to a drop in consumer demand as residents decrease, the progressive aging of the population also wears down demand (reduced food consumption, less wear and tear on shoes, etc.), all of which makes business less profitable. That being said, people cannot live without buying food and daily necessities. It is possible to visit stores at a distance if one can still drive or ride a bicycle. However, when it becomes impossible to go out independently, individuals are then forced into situations where their very survival is threatened. Even though people can occasionally visit stores with the help of neighbors and local government workers, as purchases center around foodstuffs that can be preserved, this may also have an adverse effect on eating habits and overall personal health.

On the other hand, shops that maintain a presence in an area where the very elderly predominate may stock products targeted toward this age group, which may also create "consumer refugees" out of residents of other ages.

2.4 Communities in Collapse

The changes brought about by aging and population decline are not only visible but have transformed communities. Unnoticed even by residents, there are other examples that steadily advance. One is the weakening of community, as seen in the example of Hatoyama New Town. It was originally thought that creating a sense of community in residential areas developed in the suburbs would be easy when the age, family structure, and income hierarchy of residential households was relatively homogenous. In fact, even in Hatoyama New Town, a tight community was formed if only partially. There were cases where children close in age became friends, leading to families socializing as a whole. An ideal living environment with a rich, well-maintained foundation for life and active socializing among residents was the supposed goal and assumed to last forever.

However, things did not proceed as expected. One main cause is that children no longer inherit and take over the family home as they once did. They prefer moving to the downtown areas, which were full of amenities and urban appeal, even if that meant a smaller home over the rich-in-nature but inconvenient suburb living. Increasingly, children living in central Tokyo are also moving their parents closer to them.[14] The backdrop for this is the decline in real estate prices that persisted nearly 20 years after the collapse of the bubble economy in the early 1990s, the unprecedented low interest rates allowing them to take on big housing loans, and the sizeable condominium and apartment house projects in urban areas that replaced the employee dormitories, athletic fields, and other corporate amenities on property sold off by companies struggling in the enduring recession.

The relationship between remaining residents also is not necessarily close. Suzuki, the aforementioned aging resident in Hatoyama New Town, is anxious about the future and claims he had very little to do with his neighbors until he retired, despite having lived there for over 40 years. This was not unusual for bed-town residents working in Tokyo. They would leave for work early in the morning by bus and train, and toil as corporate warriors until late at night. Each felt lucky if they could get home on the last train. If not, employees stayed over countless nights at hotels near the company. No one had the energy or the time to devote to socializing with neighbors.[15]

Upon moving into general housing developments, even in Hatoyama New Town, one was supposed to join the residents' association and for a while it remained active. However, when construction of a nearby golf course was proposed, residential opinion was split in two over the environmental impact, leading to a winnowing of cordial relationships (Yamazaki 2010). The residents' association subscription rate also plummeted, falling to around 30% today. Recently, the local festival, which has been held for almost 40 years, has become impossible without the help of town staff and students from the nearby high school. Furthermore, the rapid development and diffusion of changing social values, and the introduction of the Internet, have further accelerated community collapse. A growing consciousness that emphasizes privacy weakened ties between neighbors. The handling of personal information has become stricter; the telephone contact network has been done away with; and it is increasingly difficult to circulate the local information bulletin. However, we also cannot deny that it is now possible to communicate anywhere and anytime using e-mail and chat services on smartphones and computers. While it is now easier to stay in touch with friends far away, time to talk with neighbors is steadily eroding.

2.5 Rising Numbers of Residents over 75 in Greater Tokyo

As the population ages, medical and nursing care issues have grown even more serious. To date the ratio of elderly in a given community—in other words the number of young people supporting an aged person—was at the center of the aging debate. However, to capture the effects of aging that will overtake the metropolitan area in the future we need to focus not only on the aging rate but on the absolute number and rate of increase in the elderly as well.

Figure 2.4 provides the number and rate of increase in people over 75 years of age (the very elderly with a high risk of severe illness), by secondary medical area (areas comprising neighboring municipalities that join forces to provide general medical care) within the metropolitan area. For reference purposes, the figure also provides the names of major secondary medical area cities with an increase rate over 100%. Although the rate in the downtown area is high, the rate for suburban municipalities where married couples have purchased detached homes or condominiums is particularly high. In these areas, the absolute number of elderly people will increase. Even within the metropolitan area, it is predicted that the population of people over the age of 75 will increase by 2.54 million between 2010 and 2025.[16] An extreme shortage in the number of beds for the very elderly is forecast. Should this become a reality, people will be unable to receive adequate care and medical services in their place of residence, and will have to travel a distance for facilities and hospitals, or even move in some instances. Advanced aging will also have a significant impact on emergency medicine. There is a risk of collapse if the current mechanisms for emergency acceptance of the critically ill in Metropolitan Tokyo remain unchanged. The number of emergency dispatches in 2013 peaked at approximately 5.91 million calls nationwide, up some 107,000 from the previous year (1.8%). This is equivalent to an ambulance being dispatched every 5.3 seconds and comes to a figure of 1 in 24 people being transported.[17]

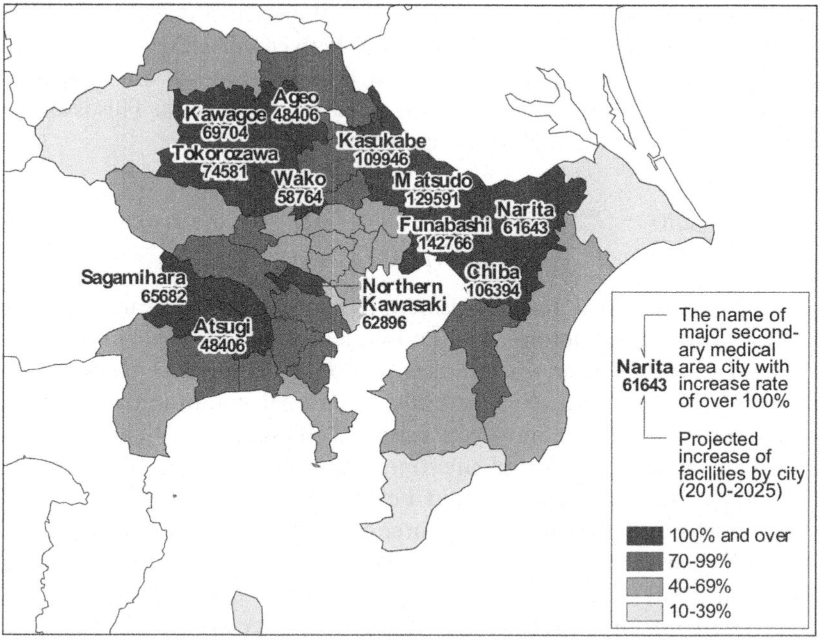

Fig. 2.4 Rate of increase in the population over 75 in Greater Tokyo. Source: MIC, National Census 2010; IPSS, Population Projections for Japan January 2012

The ratio of doctors to population has always been low in Greater Tokyo, especially Saitama, Chiba, and Kanagawa Prefectures. As a result, even if ambulances rush to the patient, finding a hospital that will accept them is not guaranteed. The share of cases where the site standby time exceeds 30 minutes was 14.6% in Saitama Prefecture, 11.2% in Chiba Prefecture, 9.6% in Metropolitan Tokyo, and 7.4% in Kanagawa Prefecture, well above the national average of 4.8%.[18] On the other hand, looking at emergency patients by age, the elderly (over 65 years of age) account for the great majority nationwide; the elderly accounted for account for 54.3% of emergency patients although they only made up 23% of the population in 2010.[19]

Under these circumstances, an increase of more than 70% in the population over 75 is predicted in 2010–2025 for many parts of Greater Tokyo. This sharp jump will make it impossible to maintain the already saturated metropolitan emergency intake system if the elderly use emergency medical care with the same frequency as they do today. Advanced aging also means an increase in residents needing assistance in the event of a disaster.[20] Furthermore, in Greater Tokyo, ties between residents in local communities are weakening and single and foreign residents are increasing, which also raises the hurdle for saving lives in emergencies. This issue will be discussed in more detail in Chapter 9.

2.6 "Marginalization" Increases Even in Central Tokyo

When a city goes into decline, residents leave in droves in search of better facilities and infrastructure. As a consequence, all who remain are those who cannot relocate or who think it is too late to move. Unsurprisingly, many of them are elderly. Currently, "marginal communities" with an elderly ratio of 50% or more are scattered in rural areas. The elderly in their familiar and rich natural environment manage to live, thanks to visits from their children, relatives, and helpers. When living at home finally becomes too difficult, they either enter city facilities and hospitals or go to live with their children, leaving the village behind. This "unsustainability" will soon reach Greater Tokyo as well. Figure 2.5 outlines the aging rate as of 2010 for Greater Tokyo, which already exceeds 25% in peripheral municipalities. Districts where the aging rate is especially marked have most likely become "marginal areas." According to the Agency of Natural Resources and Energy, a characteristic is an upside-down population pyramid, where aging is much sharper in areas where residences and social capital grew during the high growth era (1960s–1972) than in areas that did not undergo as much development.[21] The major cause for this is the sudden influx of a given cohort immediately after development and the outflow of their children, who have not taken over their parents' homes.

Looking at forecasts for the rate of demographic change in 2010–2040 for the Tama area in Tokyo, marginalization is already severe with resident numbers falling by more than half in Okutama (down 58.6%) and Hinohara (down 52.1%). However, increases even in the Tama Area are predicted for Inagi (up 9.4%), Mitaka (up 1.8%), and Higashi Murayama (up 0.7%), suggesting that life in areas outside Tokyo's 23 central wards are not necessarily unsustainable.[22] In these areas in decline, the disadvantaged who cannot move due to a lack of resources or some other reason will be left behind. Housing experts refer to this as "residualization" (Kubozono et al. 2010). In fact, residualization is already gaining ground even in Greater Tokyo. It is in public housing that this is especially noticeable.

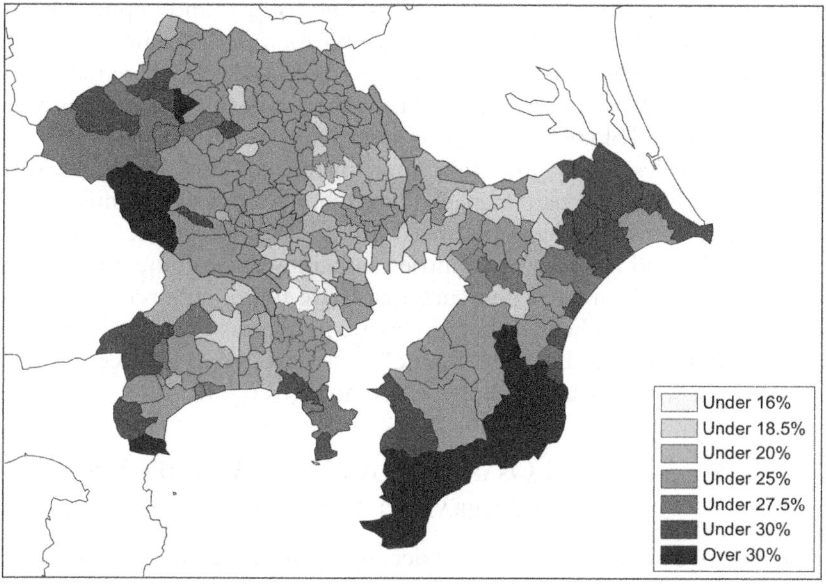

Fig. 2.5 Population aging in Greater Tokyo (2010). Source: MIC, National Census 2010

Public housing was originally developed in line with the state's policy of providing a better standard of living when there was a shortage of housing in the postwar era. Municipalities received state subsidies to build public housing and to provide inexpensive shelter for low- and medium-income citizens. It was originally assumed that public housing residents would build up their resources and eventually move out. Youths who flocked to Tokyo from the countryside in the postwar high-growth period (late 1950s–1970s) moved out of public housing as the government plan predicted. However, many also stayed behind. They were already aging, many were retired and living off pensions, or had no intention or the means to leave public housing at that stage of life. In this way, those unable to escape from impoverished circumstances—such as the elderly as well as single-parent households—remained in public housing. This phenomenon is more likely to occur in large cities like Tokyo than in regional locales with their high aging rates. The main reason is because the absolute number of the elderly is higher in large cities. There, low-cost and reasonably well-maintained public housing has become one of the few places where low-income seniors can live without worry.

This residualization in Greater Tokyo is not only found in public housing. Although the reasons and background for residualization are quite different, aging residents can be seen in rundown wooden rental apartments even in dense urban areas.[23] How, then, will this housing residualization change Tokyo? Urban areas with many inactive elderly will not only become less lively, but their atmosphere will also potentially stifle growth. Without adequate management, there is a risk of "slumification." In general, slums are residential areas with poor infrastructure, health and safety issues, and are seen in countries and cities undergoing a rapid process of growth. In many cases, such places are full of children and young people, and also display a certain liveliness. In contrast, the slums that will be seen in Tokyo will be in the "static" state depicted at the beginning of this chapter with a risk that there will be no signs of life.

2.7 Greater Osaka Becomes the World's First Degenerating Megacity

The negative impact of population decline is also reflected in metropolitan areas besides Tokyo. In the Osaka area, apart from advanced aging and a declining population, the number of people commuting by train from the suburbs to the city center is falling. Therefore, the situation could also be seen as a shrinking of the metropolitan area. According to the Kinki Person Trip Survey, conducted every 10 years on traffic movement in the Keihanshin Area (Osaka, Kyoto, and Kobe), inter-area weekday commuter trips (travel from one point to another counted as one unit) had been growing in many areas up until 2000 accompanying an expansion of the metropolitan area. In 2010, however, the number of trips between the Osaka city center and other points fell significantly (Fig. 2.6). There are several key reasons, including a rise in people living in the city center and a drop in population on the outskirts, the retirement of the generation that had commuted from the outskirts of residential areas to the city center, and a growing aversion to long commutes (Takata 2016). If this trend continues, the major railway companies that have supported travel in the Kansai region, to the extent that the area is called the "private railway lines kingdom," will be hard hit. Each company is currently pursuing cost-cutting measures such as introducing unmanned stations not only on the branch lines built accompanying suburban development but also on major routes. In the future, services may be significantly reduced and stations and lines abandoned.

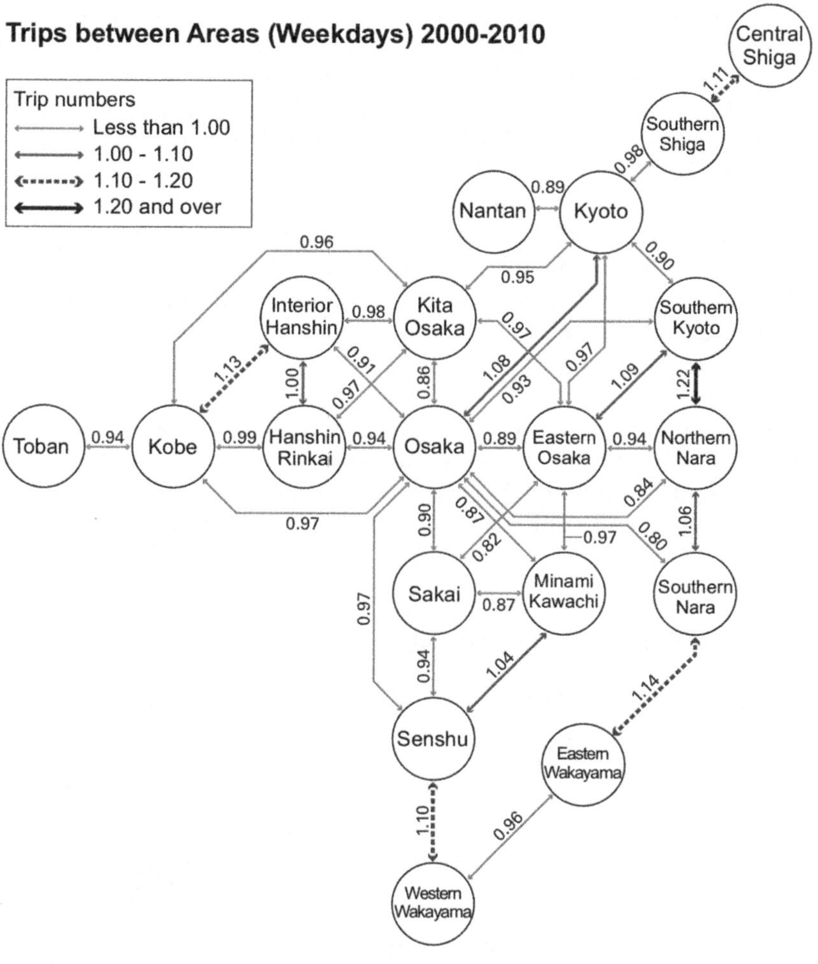

Fig. 2.6 Changes in number of trips between areas in Greater Osaka. Source: Keihansin Metropolitan Transportation Planning Council. Fifth Greater Kinki Person Trip Survey 2010. Note: The figure shows cases of approximately 50,000 trips or more

Urban residents who believed unmanned stations belonged to rural or tram lines were shocked to find themselves commuting via stations with no employees. Although automatic ticket vending machines and ticket gates are common now, the fact that station staff could be approached when in trouble provided peace of mind. Many Japanese conceived of the railway station as being akin to the police box, an official presence that watched over the community. Although residents opposed unmanned stations, the railway companies could not abandon their cost reductions. Local residents were forced to cave in thinking it was better than having no station at all.

Greater Osaka will become the world's first megacity (urban areas with a population of 10 million or more) to experience a long-term drop in its population.[24] There are several megacities in Asian countries that have undergone rapid economic growth, but population continues to grow in these cities. According to a government report, the fall in population in Greater Tokyo and Greater Nagoya is expected to proceed more slowly than the national average, but more rapidly in Greater Osaka (Fig. 2.7).[25] The most significant impact of a degenerating megacity can be seen in its fringes. There are many residential areas that sprang up just as the metropolitan population shifted from growth to decline left behind in the fringe when the population did not expand as expected. Although some people

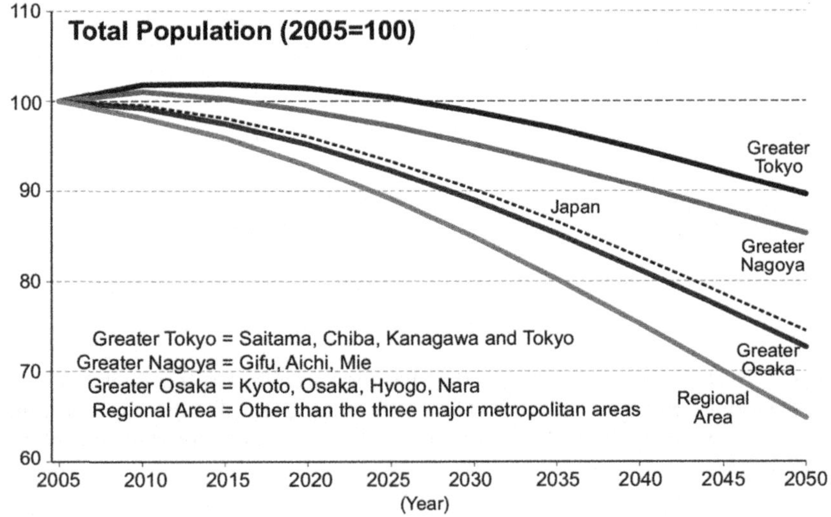

Fig. 2.7 Changes in projected population by area. Source: MLIT National Land Development Council, Policy Subcommitee for Long-Term Outlook, Interim Report on the "Long-Term Land Outlook" 2011

bought land for speculative purposes in anticipation of an increase in land prices, many others took out mortgages to actually live there. An enormous tragedy awaits these people. If people do not gather in residential areas, problems like worsening security and cityscapes arise, as well as the spread of a depressing atmosphere. The living environment deteriorates rapidly with empty houses taken over by animals and rampant illegal dumping on vacant lots. In residential areas where residents are responsible for managing sewage and streetlights and so on, the upkeep of these lifelines becomes more precarious as they decrease.

2.8 Greater Nagoya's Bottleneck—Linear Trains and Few Female Workplaces

Compared to Greater Osaka and its ongoing population decline, Greater Nagoya, one of Japan's three great urban concentrations, is expected to maintain a level of population by attracting people and industry from the surrounding Aichi Prefecture. The manufacturing industry is strong in Aichi and consumption is also robust, acting as a significant unifying force in the greater Nagoya region. On the other hand, the two neighboring prefectures of Gifu and Mie are both experiencing advanced depopulation. The decline of these prefectures has even caused concerns about a dampening of the vitality of Aichi Prefecture.

When commuting time to Greater Tokyo shortens with the opening of the Linear Bullet Train (a maglev train), this may well draw people to Greater Tokyo from Nagoya. The manufacturing industry that is Nagoya's strength generally offers more employment for men than women. Women seeking work may well shift their place of residence to Greater Tokyo.[26] The new Linear Bullet Train may be an opportunity for Nagoya companies to shift their head office functions to the capital (Okada 2013).

2.9 Sapporo and Fukuoka Absorb Their Surrounding Populations

The number of people taking up residence in Sapporo and Fukuoka during 2011–2013 far outnumbered those leaving and these were the next largest population inflows in the nation's urban areas after the 23 wards of Tokyo (Sapporo second with some 29,500; Fukuoka third with some 28,500) (Yamamoto 2014). Both attracted a large number of residents from within their respective islands of Hokkaido and Kyushu. Sapporo is characterized by a large number of elderly incoming residents. As hospitals all over Hokkaido disappear due to financial difficulties, it is believed that more and more people seeking hospital services are moving to Sapporo. On the other hand, Fukuoka is characterized by a large number of young inhabitants, which will be investigated more closely in Chapter 10.

In this chapter, we surveyed how population decline is not a phenomenon confined to the regions but is increasingly being felt in Japan's major metropolises. It is important to understand how changes in the population are affecting regional infrastructure as these issues will likely affect the urban centers, as will be discussed in Chapter 3.

Notes

1. MILT 2015; MIC Statistics 2016.
2. Suzuki, Taro, personal communication, December 2014.
3. Ibid.
4. Yamazaki 2010.
5. Ibid.
6. Ibid.
7. Nihonshintoshikaihatsu Corporation.
8. IPSS 2012.
9. Nishioka et al. 2011; IPSS 2013.
10. Choshi City Government 2008; MIC 2011; IPSS 2013.
11. Tokyo Metropolitan Government 2015.
12. MLIT 2015.
13. Utsunomiya 2013.
14. Cabinet Office 2011.
15. Suzuki, Taro, personal communication, December 2014.
16. This calculation is based on data from the National Institute of Population and Social Security Research (IPSS 2013).
17. Fire and Disaster Management Agency 2014.
18. MIC 2010.
19. Fire and Disaster Management Agency 2014.
20. MIC 2013.
21. Cabinet Office 2011.
22. IPSS 2013.
23. Tokyo Metropolitan Government 2012.
24. Osaka Prefectural Government 2014.
25. MLIT 2011.
26. Central Japan Railway Company 2012.

References

Bureau of Urban Development, Tokyo Metropolitan Government. (n.d.). The appeal of Tama New Town (Tama nyutaun no miryokutte). http://www.toshiseibi.metro.tokyo.jp/bosai/tama/miryoku.html, accessed on September 21, 2016 (in Japanese).

Cabinet Office, Government of Japan. (2011). Local economy 2011 (Chiiki no keizai 2011). http://www5.cao.go.jp/j-j/cr/cr11/chr11040202.html, accessed on September 21, 2016 (in Japanese).

Central Japan Railway Company. (2012). JR Tokai annual report 2012. http://english.jr-central.co.jp/company/ir/annualreport/_pdf/annualreport2012-05.pdf, accessed on September 21, 2016.

Choshi City Government. (2008). Closure notice of the Choshi City Hospital (Choshi shiritsu sogobyoin kyushi no oshirase). http://www.city.choshi.chiba.jp/sisei/siritubyouin/byouinkyushinoshicyou.html, accessed on September 21, 2016 (in Japanese).

Fire and Disaster Management Agency. (2014). Fire and disaster white paper (Shobo hakusho). http://www.fdma.go.jp/html/hakusho/h26/h26/, accessed on 8 February 2017 (in Japanese).

IPSS. (2012). Population projections for Japan (January 2012): 2011 to 2060. http://www.ipss.go.jp/site-ad/index_english/esuikei/ppfj2012.pdf, accessed on September 21, 2016.

IPSS. (2013). Regional population projections for Japan: 2010–2040. (Nihon no chiiki betsu shorai suikei jinko). *Population Research Series* No. 330, http://www.ipss.go.jp/pp-shicyoson/j/shicyoson13/6houkoku/houkoku.pdf, accessed on February 8, 2017 (in Japanese).

Keihansin Metropolitan Transportation Planning Council. (2010). Fifth Greater Kinki Person Trip Survey (Dai go kai kinkiken pason torippu chosa). https://www.kkr.mlit.go.jp/plan/pt/data/pt_h22/index.html, accessed on February 14, 2017 (in Japanese).

Kidachi, Manao. (2011, November). What are food deserts? *Yomiuri Shimbun*. http://www.yomiuri.co.jp/adv/chuo/dy/opinion/20101101.html, accessed on September 21, 2016.

Kubozono, Yoichi, Fumihiko Seta & Yahagi, Hiroshi. (2010). Study on public housing residuary (Koeijutaku no zanryuka ni kansuru kenkyu). *Journal of Architecture and Planning* 75, no. 649, 659–666. https://www.jstage.jst.go.jp/article/aija/75/649/75_649_659/_pdf, accessed on September 21, 2016 (in Japanese).

MIC. (2010). Emergency dispatches in 2010 (Heisei 22nen kyukyuhanso ni okeru iryokikan ukeire chosakekka). http://www.fdma.go.jp/neuter/topics/houdou/h23/2307/230722_1houdou/03_houdoushiryou.pdf, accessed on September 21, 2016 (in Japanese).

MIC. (2011). National Census in 2010 (Heisei 22nen kokusei chosa). http://www.stat.go.jp/data/kokusei/2010/kihon1/pdf/gaiyou1.pdf, accessed on February 14, 2017 (in Japanese).
MIC. (2013). Guidelines of evacuation support for people who need assistance in emergency (Kinkyukodo yoshien no hinankodo shien ni kansuru torikumi shien). http://www.bousai.go.jp/taisaku/hisaisyagyousei/youengosya/h25/pdf/hinansien-honbun.pdf, accessed on September 21, 2016 (in Japanese).
MIC Statistics Bureau. (2016). Outline of the 2015 population census of Japan (Heisei 28nen kokusei chosa kekkano gaiyo). http://www.stat.go.jp/data/kokusei/2015/kekka/pdf/gaiyou.pdf, accessed on September 21, 2016 (in Japanese).
MLIT. (2011). Long term vision of land management-mid report (Heisei 17 nendo Kokudo no chokitenbo, chukan torimatome). http://www.mlit.go.jp/policy/shingikai/kokudo03_sg_000030.html, accessed on September 21, 2016 (in Japanese).
MLIT. (2013). Promoting rebuilding old buildings (Rokyuka mansion no tatekae tou no sokushin nit suite). http://www8.cao.go.jp/kisei-kaikaku/kaigi/meeting/2013/committee2/131024/item1-1.pdf, accessed on February 14, 2017 (in Japanese).
MLIT. (2015). White paper on Land, Infrastructure and Tranport 2015 (Kokudo Kotsu Hakusho 2015 nen). http://www.mlit.go.jp/hakusyo/mlit/h26/hakusho/h27/pdf/np101200.pdf, accessed on September 21, 2016 (in Japanese).
Nishioka, Hachiro, Shiro Koike, Masakazu Yamauchi, Keita Suga & Yuji Esaki. (2011). Population projections by prefecture in Japan: 2005–2035: Outline of results and methods. *The Japanese Journal of Population*, 9(1), 1–39.
Okada, Yutaka. (2013). *Significant differences in the future for the three major metropolitan areas (Sandai toshiken shoraizo ni okina chigai)*. http://www.mizuho-ri.co.jp/publication/research/pdf/insight/pl130930a.pdf, accessed on September 21, 2016 (in Japanese).
Osaka Prefectural Government. (2014). White paper on declining population trend of Osaka prefecture (Osakafu jinko gensho shakai hakusho). http://www.pref.osaka.lg.jp/attach/10938/00122693/tyouryuu_26.6.pdf, accessed on September 21, 2016 (in Japanese).
Takata, Hajime. (2016, April 20). National Census—depopulation society and population concentration (Kokusei chosa ga shimesu jinko gensho to toshi shuchu no genjyo). http://www.mizuho-ri.co.jp/publication/research/pdf/today/rt160420.pdf, accessed on September 21, 2016 (in Japanese).
Tokyo Metropolitan Government. (2012). Plan for 10 year fireproof acceleration project for wooden housing area ("Kimitsu chiiki funenka 10 nen purojekuto" jishi hoshin). http://www.metro.tokyo.jp/INET/KEIKAKU/2012/01/DATA/70m1k100.pdf, accessed on February 14, 2017 (in Japanese).

Tokyo Metropolitan Government. (2015). Tokyo plan 2015 (Tokyo-to keikaku 2015). http://www.fukushihoken.metro.tokyo.jp/kourei/shisaku/kyo-jyuuannteikakuho/kakuhoplankaitei.files/zenbun.pdf, accessed on September 21, 2016 (in Japanese).

Utsunomiya, Toru. (2013, April 4). Is Seibu Chichibu line really unable to make profit? (Seibu chichibu senha hontoni shuekiryokuga nainoka). *Toyo Keizai Online*. http://toyokeizai.net/articles/-/13557, accessed on October 6, 2016 (in Japanese).

Yamamoto, Kenzo. (2014). *Why are Sapporo and Fukuoka in the top three population influx? (Sapporo, Fukuoka wa naze jinko ryunyu toppu 3 nanoka)*. http://www.keieiken.co.jp/pub/yamamoto/column/column_140203.html, accessed on September 21, 2016 (in Japanese).

Yamazaki, Tsutomu. (2010). *No future for Japanese property (Fudosan zetsubo mirai)*. Tokyo: Toyo Keizai Shinposha (in Japanese).

Yoneyama, Hidetaka. (2012). Future prospects for vacancy rates and countermeasures (Akiya ritsu no shorai tenbo to akiya taisaku). *Fujitsu Research Institute research report*, no. 392. http://jp.fujitsu.com/group/fri/downloads/report/research/2012/no392.pdf, accessed on September 21, 2016 (in Japanese).

CHAPTER 3

A Collapse in Regional Infrastructure

Igarashi Chikako, Akiyama Yuki, and Kamiya Kenichi

Japanese authorities are concerned that a declining population will exacerbate a collapse in regional infrastructure. If regional depopulation advances at the current pace, the provision of utilities such as water and gas, as well as public transport, will decline precipitously, and maintaining such services may be difficult. Moreover, if customer numbers keep falling, shops and gas stations may be forced to close. The risk of falling into this predicament is higher in regional areas, and it is also conceivable that daily life will become untenable. In order to prevent such a fate, each region needs to design its own future and begin taking measures immediately.

3.1 Expanding "Uninhabited Zones"

The question of population decline is not just a matter for depopulated areas. Residents will certainly decrease in wide swaths across the country, especially in rural regions. In 2014, the Ministry of Land, Infrastructure,

Igarashi. C
General Incorporated Association Hokkaido Intellect Tank, Sapporo, Japan

Akiyama. Y (✉)
Center for Spatial Information Science, University of Tokyo, Tokyo, Japan

Kamiya. K
Medical Care Research Department, General Incorporated Association Hokkaido Intellect Tank, Sapporo, Japan

Y. Funabashi (ed.), *Japan's Population Implosion*,
https://doi.org/10.1007/978-981-10-4983-5_3

Transport and Tourism (MLIT) announced its *Grand Design of National Spatial Development Toward 2050, Japan—Creation of a Country Generating Diverse Synergies Among Regions.* This provided a blueprint for regional development in the medium and long term on the basis of population estimates for 2050. The ministry proposed sustainable community building where the population is in decline by securing places of employment and essential services for daily life, such as shopping, hospital and medical care, and elderly care service. The ministry divided the country into one square kilometer units and indexed the long-term population projections for each sector. Based on these statistics, officials calculated that nearly 20% of currently inhabited areas would become "uninhabited zones" by 2050 (Fig. 3.1).[1] Should these zones be scattered throughout a given region, it would become difficult to maintain infrastructure necessary for daily living, first in rural areas, but as population decline advances, it would then also soon spread to urban areas.

What will happen to this infrastructure? Below we make several specific predictions.

3.2 The Reality of a Shrinking Population

In 2008, Japan's population as a whole began to plummet. In rural areas, however, depopulation was already on the rise. But several prefectures experienced an increase and a decrease from the 1980s to the 2000s—11 prefectures peaked before 1989, 10 before 1998, and 16 before 2008. Prefectures that peaked early were concentrated in the northeastern Tohoku region, the central Chugoku region and the islands of Shikoku and Kyushu.

In the past, Japan's population decline had progressed relatively slowly. In the five prefectures with the highest rate of decrease between 1985 and 2010, there was a less than 10% drop. However, around 2020 to 2030 the population decline is expected to advance more rapidly. Consequently, most people—especially in urban areas—will experience a real sense of population decline from around 2030. According to the 2013 *Japan's Regional Population Projections*, announced by the National Institute of Population and Social Security Research (IPSS), the rate of decline in 42 prefectures out of 47 will be more than 10% between 2010 and 2040.[2] Moreover, even those prefectures currently experiencing population growth will shift into decline, even in the last remaining prefecture of Okinawa between 2020 and 2025. As a result, by 2060 the population in most prefectures will be 50–70% of its 2010 level.

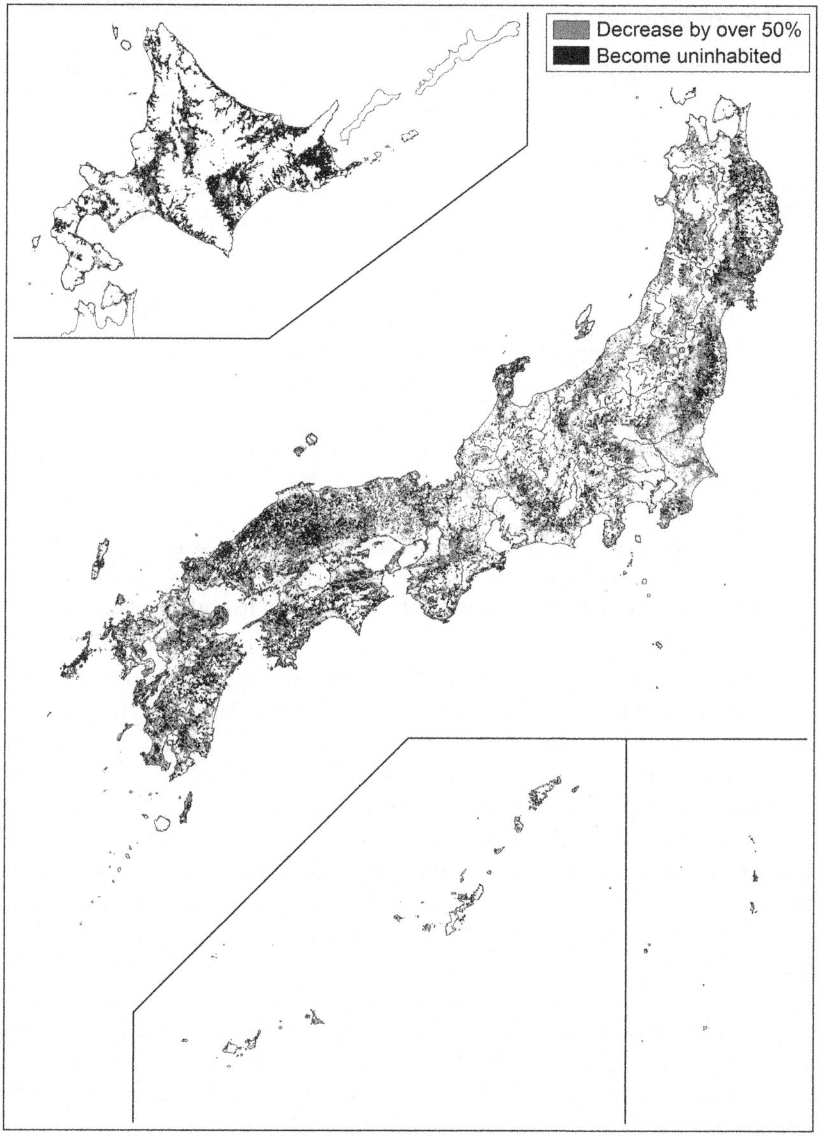

Fig. 3.1 State of depopulation in 2050 (National 1 km mesh) (2010 = 100). Source: Compiled from MLIT, Reference material of Grand Design of National Spatial Development towards 2050, July 2014

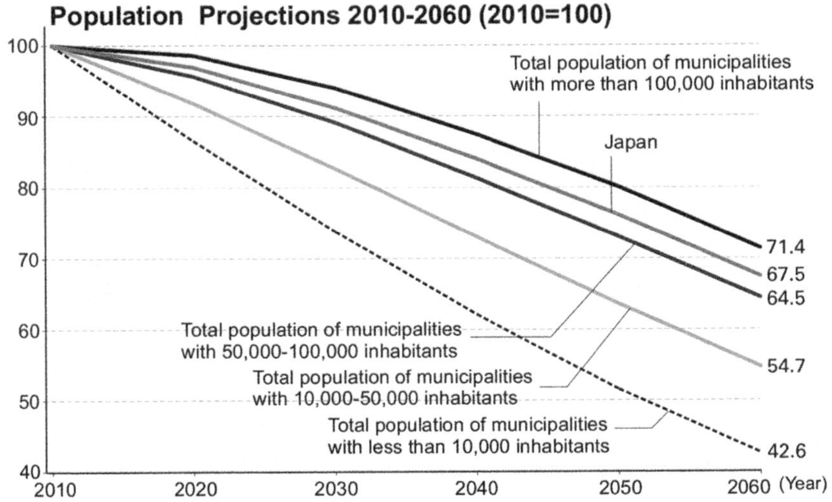

Fig. 3.2 Changes in the depopulation rate by city scale. Source: IPSS, Japan's Regional Population Projections, March 2013. Note: Figures after 2040 were extrapolations calculated by the authors from the same hypothetical figure

Looking at individual municipalities, the differences are even starker. Those estimated to have a higher population in 2040 than in 2010 account for only about 5% of the nation, whereas some 70% of all municipalities expect to see a more than 20% reduction in their populations. Furthermore, by 2060, the population is expected to be less than or equal to half of the current level in approximately half of the municipalities nationwide.[3] The smaller the size of the municipality, the faster the speed of population decline. Towns and villages with 10,000 people or less at present are estimated to be approximately 40% of the current figures by 2060 (Fig. 3.2).

There are several factors behind the decline in regional population, including "a drop in migration of population to other areas [and] a drop in the birth rate." After World War II, a large number of young people left their local communities for the big cities. In the high growth era of the 1960s, more than 600,000 junior high graduates migrated en masse every years to the three major metropolitan areas of Tokyo, Osaka, and Nagoya in search of employment opportunities. As a result, local communities lost

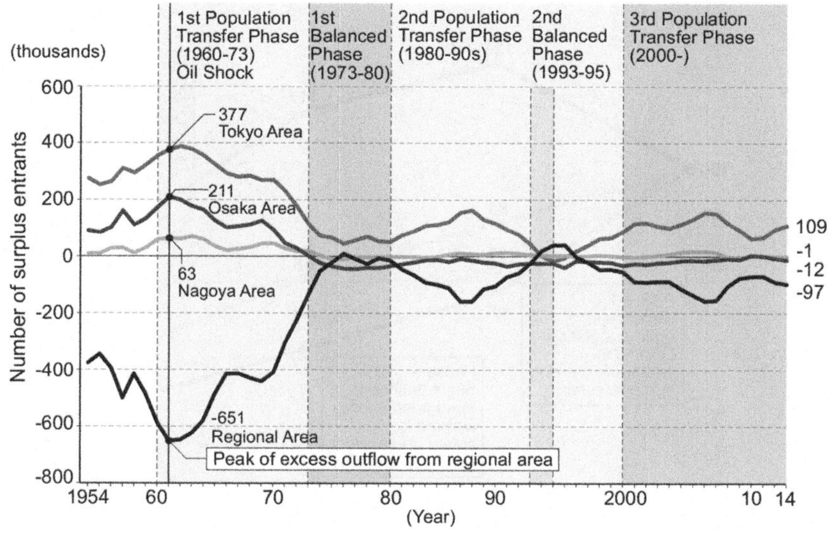

Fig. 3.3 Changes in population transfers between the three major metropolitan areas and the regional area. Source: MIC, Statistics Bureau, Report on Internal Migration in Japan

a large number of youths who would have helped to maintain population levels. In recent years, after the large-scale population movements from the countryside to large urban areas during the bubble economy years of the 1980s, there was a greater tendency for people to move from the regions to Greater Tokyo alone (Fig. 3.3).

One factor behind this centripetal pull is the over-concentration of large companies in Tokyo. According to the Ministry of Internal Affairs and Communications (MIC) Statistics Bureau, there are 1.8 million companies and 2.8 million establishments (corporate headquarters and branch offices, sales offices, stores, etc.) with a combined total of approximately 41 million employees in Japan. Of these, 15.5% of companies, 21.7% of establishments, and nearly 30% of employees are concentrated in Tokyo. Looking only at those large firms with capital reserves of 300 million yen or more, the degree of concentration in Tokyo is 42.9% of companies, 54.7% of offices, and 52.7% of employees. Because Tokyo's population accounts for about 10% of the entire country, the degree of concentration of large corporate employees is quite high.[4]

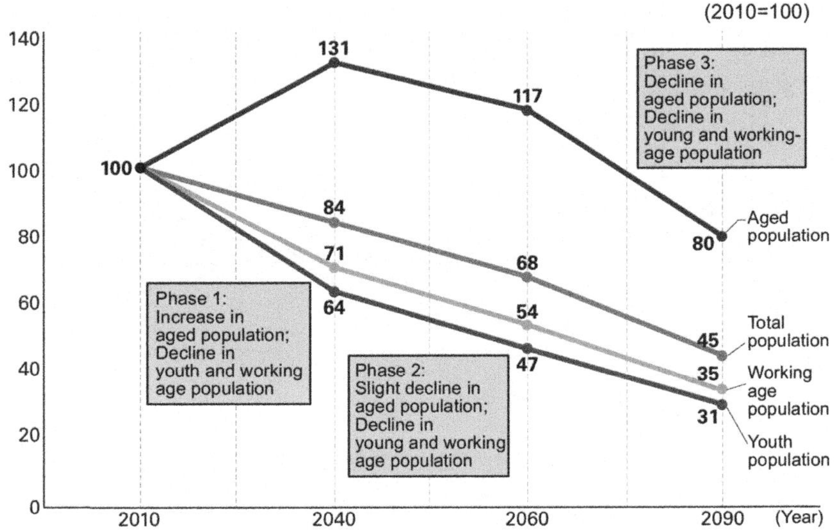

Fig. 3.4 Population projections by age grouping. Source: Cabinet Office, Headquarters for Overcoming Decline and Vitalizing Local Economy in Japan, Japan's Long-term Vision and Comprehensive Strategy, 2014; IPSS, Population Projections for Japan, January 2012

On the other hand, although there are differences in the birth rates between prefectures and nationwide, the downward trend of the population continues, and the total fertility rate (TFR) bottomed out in 2005. While the TFR is now rising, it is still far from the population replacement level required (2.07) and the number of births remains low and will continue to decrease. Excluding the three major metropolitan areas, the TFR in 2013 was lowest in Hokkaido at 1.28, followed by Miyagi Prefecture at 1.34 and Akita Prefecture at 1.35. Conversely, it was highest in Okinawa Prefecture at 1.94, followed by Miyazaki Prefecture at 1.72, and Shimane Prefecture and Kumamoto Prefecture tied at 1.65.[5] This data shows a general tendency for high birth rates in western Japan and low birth rates in northeastern Japan. In addition to this drop in birth rates, the elderly population will start to fall from 2040, spurring a further population decline (Fig. 3.4). While the term "demographic aging" gives the impression that the elderly population is increasing, the reality is actually different. A drop in the aging population is predicted from 2010 to 2040 in the three prefectures of Akita, Shimane, and Kochi. At the municipal level, the elderly population is expected to fall in 757 municipalities nationwide (45% of all municipalities), especially in those with small populations (IPSS 2013a).

3.3 Minimum Population for Maintaining the Life Infrastructure

The MLIT's *Grand Design of National Spatial Development Toward 2050* analyzed the results and published the population size figures for "a municipality where the probability for service facilities being located would be 50% and 80% (excluding the three major metropolitan areas)."[6] These estimates indicate the scale of a municipal population required for a 50–80% probability of private-sector services (restaurants, general clinics, cram schools, etc.) being able to thrive. For example, there is a 50% probability for the presence of books and stationery retail in a municipality of 1500, and an 80% probability for a municipality of 2500. According to this analysis, even in small municipalities with 1000 residents or less, necessary services such as food and beverage retailers, post offices, clinics, and eldercare welfare facilities will exist close by. However, services like banks, elder day care and short-stay facilities require a minimum population of 6500 people; men's clothing retail and long-term elder care facilities necessitate at least 9500 people in a community to be viable; and shopping centers, fitness clubs, and financial transaction businesses only can be found in cities of 40,000–80,000 people. A population of approximately 20,000 inhabitants is necessary for a 50% chance of having corporate services, such as tax accountant offices and rental cars, nearby. Businesses in municipalities that fall below these levels will have to rely on suppliers in other municipalities to gain equivalent services. Additionally, when a community population exceeds 20,000, the probability of corporate services along with necessities such as retail shopping, finance, education and learning support surpasses 50%, offering adequate social infrastructure.

According to MLIT's 2050 projections, the number of municipalities with 2000 residents or less will increase sharply to 166 (it was already 61 in 2010), and those with a population of 2000–4000 will rise to 182 (from 98 in 2010). These estimates suggest that the number of municipalities lacking various service facilities to support residents' daily needs will increase by around 2050 (Fig. 3.5).

	Population size with 50% probability	Population size with 80% probability
Restaurants, etc.	1,000 or less	
Food/drink retail	1,000 or less	
Post office	1,000 or less	
General medical clinic	1,000 or less	
Dentist	1,000 or less	2,001–3,000
Elder care welfare facility	1,000 or less	4,001–5,000
Books/stationary retail	1,001–2,000	2,001–3,000
Cram school	1,001–6,000	6,001–7,000
Bank	5,001–7,000	9,001–10,000
Daycare/short stay care	6,001–7,000	9,001–10,000
Long-term health care facility	9,001–10,000	20,001–25,000
Men's apparel retail	9,001–10,000	20,001–25,000
Tax accountant office	17,001–18,000	25,001–30,000
Car rental	20,001–25,000	50,001–55,000
Fitness club	40,001–45,000	60,001–65,000
Financial transactions	50,001–55,000	85,001–90,000
Shopping center	75,001–80,000	

Fig. 3.5 The population size for municipalities where the probability of locating service facilities would be 50 and 80% (excluding the three major metropolitan areas: Tokyo, Chiba, Saitama, Kanagawa; Aichi, Gifu, Mie; Osaka, Kyoto, Hyogo, and Nara). Source: Compiled from MLIT, National Spatial Planning and Regional Policy Bureau documents

3.4 Disappearing Life Infrastructure

Population decline will also change the shape of Japan's regions landscape, as illustrated by a few simple examples below.

3.4.1 Supermarkets

Of all daily services, supermarkets, which sell food, clothing and other daily necessities, are most closely connected to our day-to-day lives.

However, a growing number of small-scale supermarkets in residential areas are now being forced to close because the car has become the main form of transport in peripheral areas. In regions where there is close to a one car per person ratio, the Automobile Inspection and Registration Information Association (AIRIA) reports that there is a growing number of people who drive to suburban shopping centers with their diverse assortment of products even for their regular shopping.[7] By 2040, however, it is expected that the number of elderly not driving will increase, adding to the importance of nearby small- and medium-sized supermarkets as places to purchase food and daily necessities.[8]

In 2010, there were approximately 20,000 small- and medium-sized supermarket outlets across the country.[9] Our analysis[10] assumes there is a possibility of closure of these stores if the population in a five-kilometer trading radius drops under 4500.[11] Figure 3.6 provides the results for the probability of closure by prefecture based on trading population projections for 2040. There were only two prefectures in 2010 that failed to meet the requirements and with more than a 10% probability that small-scale supermarkets would close: Shimane Prefecture (16%) and Kochi Prefecture (13%). This will increase to nine prefectures in 2040 (Hokkaido, Iwate, Akita, Tottori, Shimane, Kochi, Nagasaki, Oita, and Kagoshima).

On the other hand, the population growth rate for those living outside the trading areas of small- and medium-sized supermarkets is high in Kochi (13.5%), Akita (9.9%), and Kagoshima prefectures (8.1%). This suggests that outlets with unviable trading areas will increase in regions where population decline begins early and will drop sharply in the future, as well as areas where residents are dispersed over a wide area. Depopulation will make it impossible to maintain the requisite number of inhabitants for a workable community of shops, leading to a vicious cycle of further commercial withdrawal. This will, in turn further accelerate population decline in the same areas. In reality, not all supermarkets will close immediately once the trading population drops below 4500. If we randomly subtract approximately half of the outlets we assume will withdraw, we can further investigate changes in the trading population for the remaining stores by municipal population size.

Surprisingly, the percentage of the population living outside of trading areas is predicted to increase by only five or so percentage points in municipal populations of 10,000 or less (Fig. 3.7). In other words, the number of residents inconvenienced as a result is unlikely to increase significantly

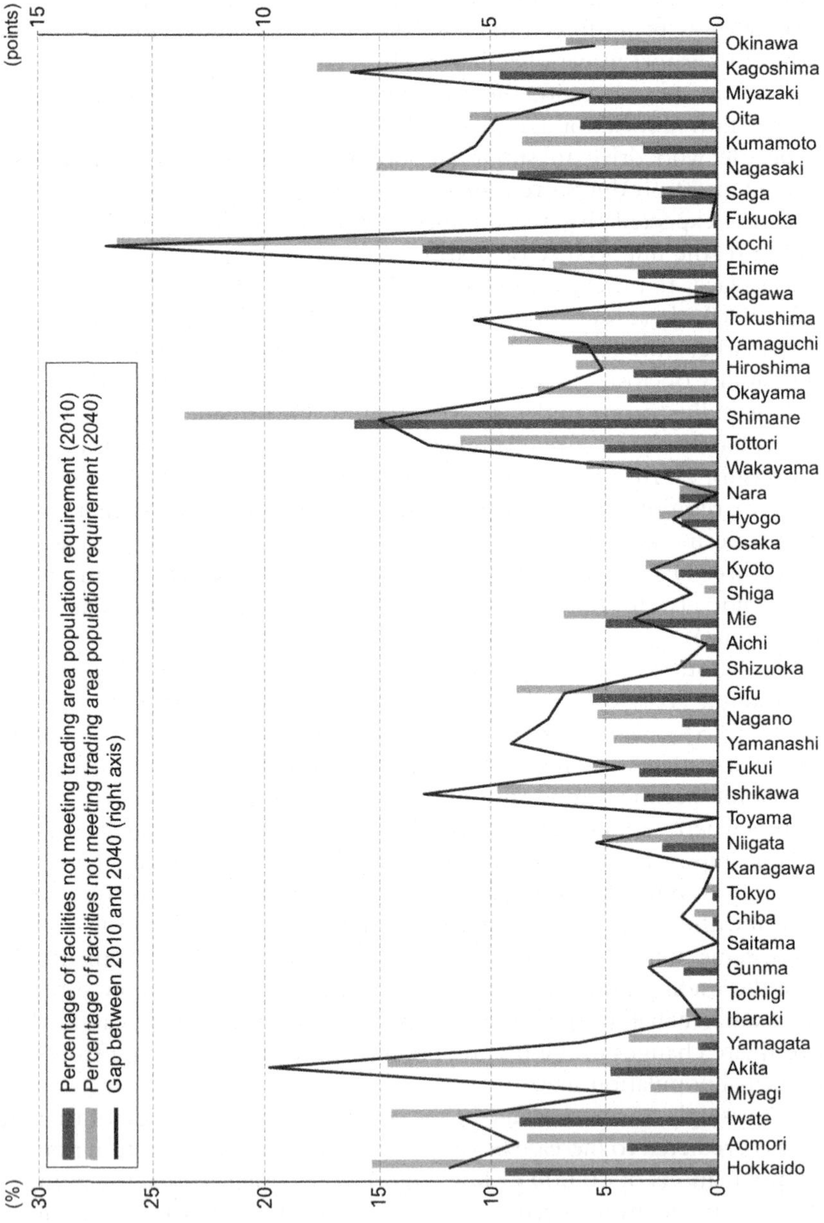

Fig. 3.6 Small and medium-sized supermarkets. Source: Joint research by CSIS and HIT

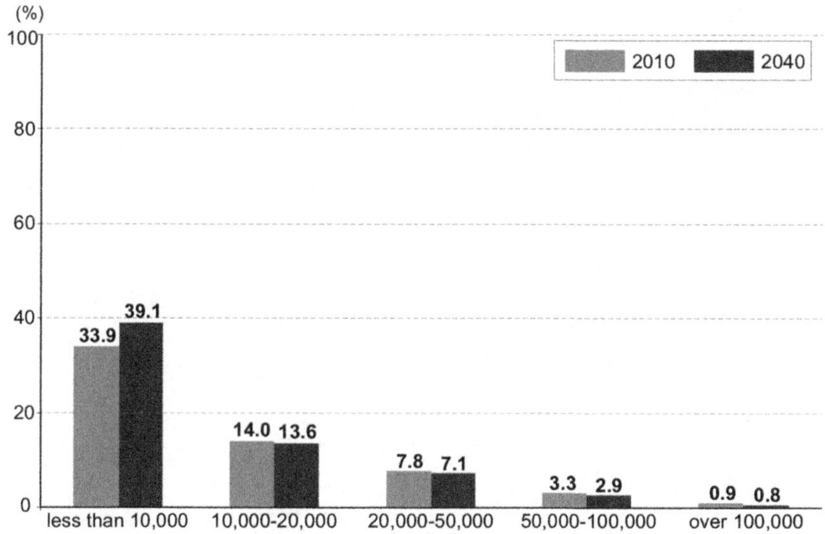

Fig. 3.7 Percentage of population living outside the small and medium-sized supermarket's trading areas (estimates for 2010 and 2040). Source: Joint research by CSIS and HIT

because supermarkets close as the ratio of population declines. Figure 3.8 provides one example of the predicted change in and around the city of Mito in Ibaraki Prefecture.[12] In some cases, residents invested funds and operate their own stores after the main supermarket closed. In 1997, residents in the Omiya-cho Tsuneyoshi district of Kyotango City, Kyoto Prefecture, opened the Tsuneyoshi Department Store. As well as selling fresh vegetables and preserved foods, the shop provides a place for children to play and the elderly to gather. Although it temporarily closed at one point, it resumed business in 2012, and has become a focal point for the local community. This trend is increasing in sparsely populated areas of Japan.

3.4.2 Convenience Stores

Convenience stores continue to expand in number across Japan. According to the Japan Franchise Association convenience store statistics, there were more than 50,000 outlets across the country in January 2015, far surpass-

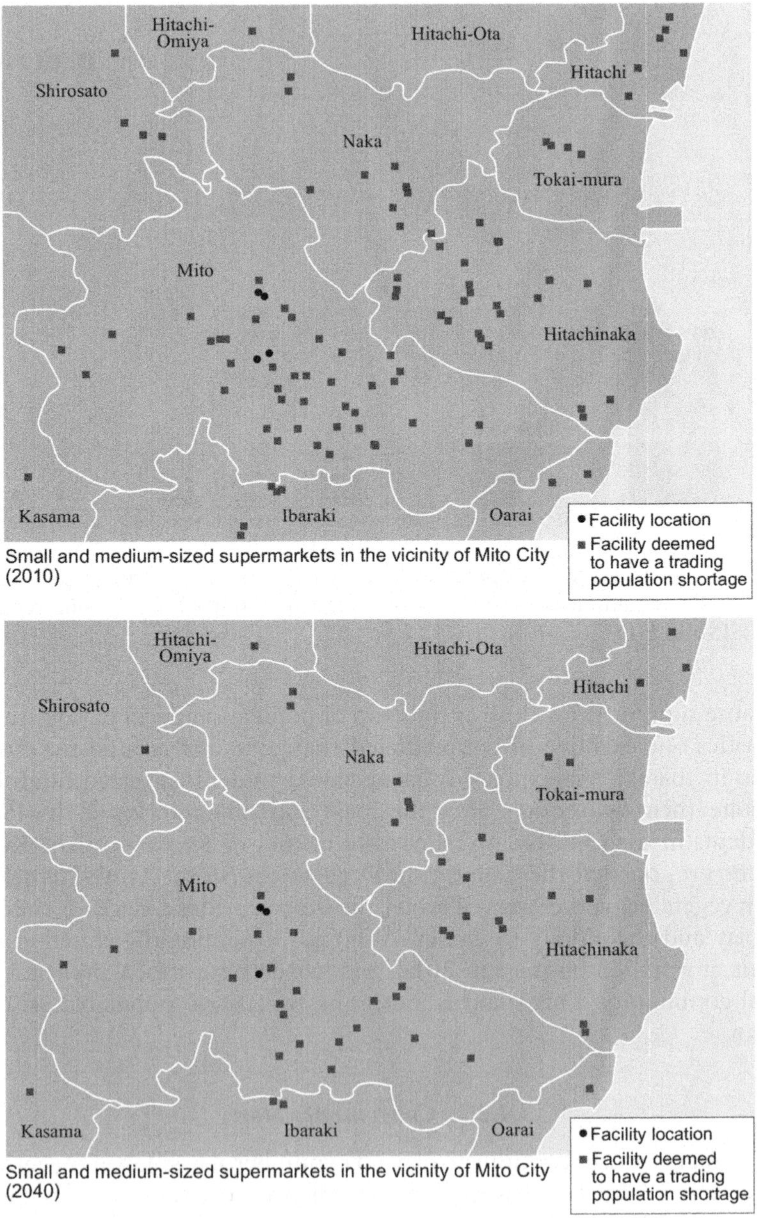

Fig. 3.8 Example of change in the number of small and medium-sized supermarkets (2010–2040). Source: Digital phone directory

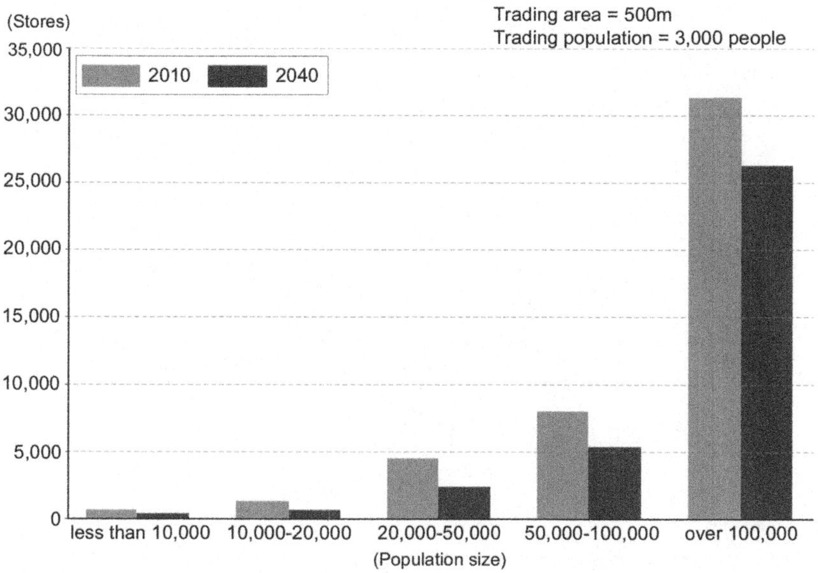

Fig. 3.9 Number of convenience stores by population size (estimates for 2010 and 2040). Source: Joint research by CSIS and HIT

ing the number of supermarkets.[13] In urban settings, convenience store trading areas are small and in the range of a 5–10-minute walk, or 500-meter radius. On the other hand, in rural areas the range augments to 10 to 15 minutes by car. The main customer base varies from company employees in business districts, to local residents in the suburbs, and tourists. The population levels necessary for them to thrive is at least 3000 in a trading radius of 500 meters, and it is estimated that more than 95% of convenience stores nationwide met this condition in 2010.[14]

As with supermarkets, the number of convenience stores is also estimated to drop to about 35,000 nationwide according to population projections for 2040 and an assumption that half will close if their trading population scale is not met. Of these, approximately 75% or 26,000 outlets will be concentrated in cities of more than 100,000 (Fig. 3.9). While the rate of concentration was about 68% in 2010, this figure will increase in the future. Figure 3.10 provides the predicted change in the number of convenience store facilities.

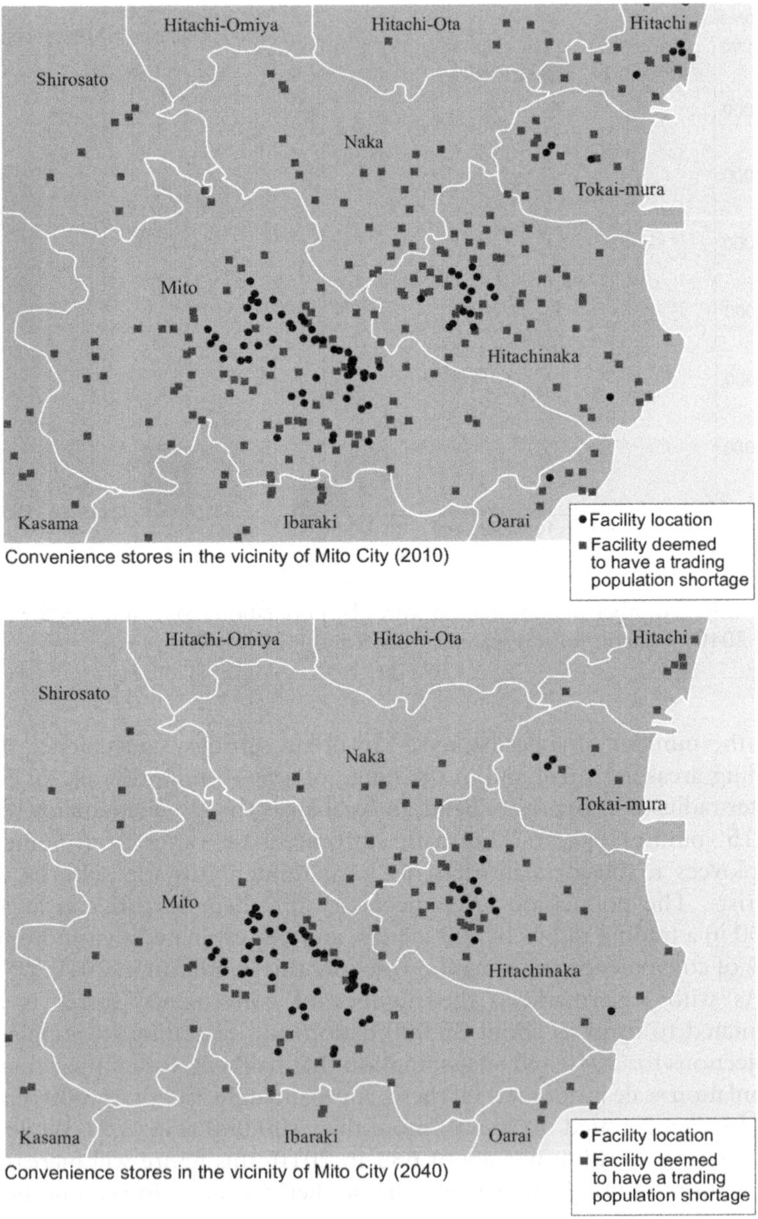

Fig. 3.10 Example of change in the number of convenience stores (2010–2040). Source: Digital phone directory

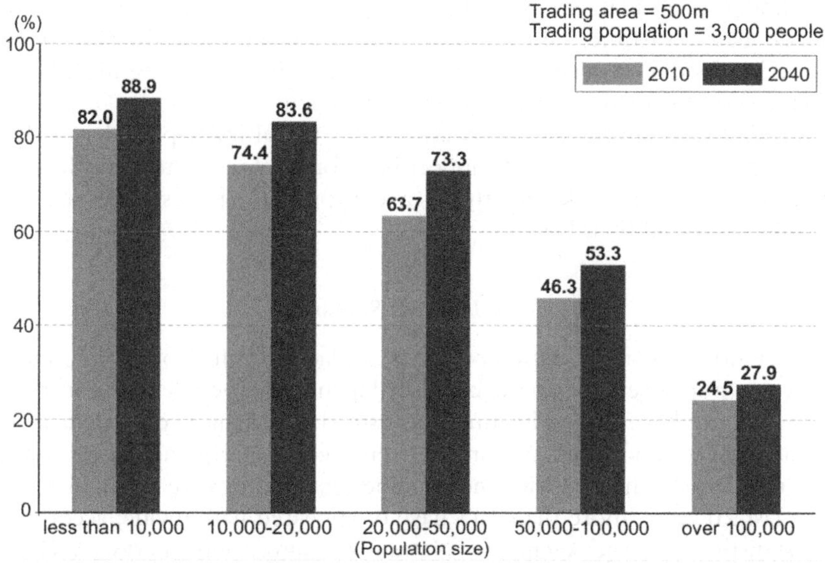

Fig. 3.11 Percentage of population living outside the convenience store's trading areas (estimates for 2010 and 2040). Source: Joint research by CSIS and HIT

On the other hand, estimates for people living outside a convenience store trading area show only a 3% increase in populations of 100,000 or more, but a 7% increase in populations of less than 10,000 and a 9% increase in populations of 10,000 to 20,000. This means the smaller the population, the more people left outside a given trading area (Fig. 3.11). While a convenience store was originally a nearby and useful place to shop for busy people in urban areas, in the peripheral or outlying regions there are a growing number of stores now catering to local residential needs by carrying fresh food and small prepared food packs, lunch boxes with local produce, hotpot cooking and a variety of sweets. Convenience stores are now not only merely convenient; they have become an essential part of life for the elderly and for people living alone. Many Japanese use these stores to pay public utility charges, use ATMs, and to pick up or drop off parcels. In the future, however, it is expected that some areas will emerge where these facilities completely shut down. Additionally, in areas where convenience stores have already withdrawn or gone out of business, the major reason seems not to be financial difficulties caused by population decline, but an inability to find a manager to take over the business or inadequate part-time staff. It is not uncommon, especially in rural areas, for a local shop owner to have obtained a franchise, meaning most likely that a shortage of successors is a serious problem.

Today, the convenience store has become the ultimate provider of "life infrastructure" and is the last support base for local residents' daily life. Its withdrawal would link directly to a collapse in services necessary for residents to lead their lives.

3.4.3 Gas Stations

Gas stations meet two different needs: a place to refuel one's car, and a place to purchase kerosene because in Japan many people living in rural areas do not have central heating but instead use smaller kerosene heaters that warm up one room at a time. Given the high frequency of car usage in rural areas compared with the big cities, gas stations are essential. Over the last 20 years, more than 40% have disappeared nationwide; according to statistics from the Agency for Natural Resources and Energy (ANRE), there were 60,000 gas stations throughout the country in 1994.[15] Since then, however, 1000 to 2000 outlets have gone out of business or shut their doors every year. As of the end of 2013, there were approximately 34,000 outlets. Gas station locations have a high correlation to population density, and there is a strong fear that they will disappear from sparsely populated areas. In 2014, there were eight towns or villages across the country with no gas stations and 63 with only one, mostly small towns and villages with a population of 5000 people or less. The impact on daily life is especially severe in tiny villages of several hundred people. There are some cases where local residents have pooled their funds to maintain an outlet. In the Kawane district in the mountains of Akitakata City in Hiroshima Prefecture, the only remaining gas station belonging to the Japan Agricultural Cooperative was set to close in 1999. Several alarmed residents launched a promotion council soliciting an investment of 1000 yen from each local household, and they now operate a gas station and grocery store. This protects the residents' daily life routine.[16]

3.4.4 Care Services

Elderly care services provided by long-term care insurance are put in place according to a Long-term Care Insurance Business Plan, drawn up every

3 years by local governments. The plan is based on the number of elderly requiring care and their needs. Currently, an "integrated community care system" is being developed for when the baby boomers enter the cohort of 75 years of age or older around 2025, usually requiring significantly more health assistance. In the future, there is a high chance that there will be a shortage in supply of necessary care services based in the home day care, home-visit care, short-term admission for daily life care, community-based services in areas with increasing elderly populations. On the other hand, care services will start to decrease especially in rural areas where the number of elderly is dropping. Multi-functional and comprehensive health centers currently being pursued in depopulated areas will be required in the future.

The debate on an Integrated Community Care System for the elderly began with a concept of providing medical care, nursing care, preventative care and a variety of life support services within 30 minutes of their residences.[17] However, not only is the 30-minute threshold not sustainable in depopulated areas, there are also places where the services cannot be set up in the first place. It is more realistic to provide a full range of services at one site or to connect to professional services in depopulated areas, positioned as "small-scale, multi-functional, composite type" services. At the same time, there is a strong possibility that the elderly will move to urban areas. There are many examples of single, elderly people moving to serviced housing in urban areas. Additionally, when a higher level of care is required and it becomes difficult for the elderly to stay in their own homes, they are sometimes admitted to nursing homes in cities. If the elderly continue to move out, there is even a possibility that a surplus in elderly and long-term care facilities developed to serve the current aging population may arise.

3.4.5 Educational Institutions

3.4.5.1 Elementary and Junior High Schools

Elementary and junior high schools are the hub of community events, and are felt to be symbols of regional economic fitness for the Japanese. Regions have been forced one after another to close these schools because of a drop in student numbers due to the declining birth rate. According to the School Basic Survey by the Ministry of Education, Culture, Sports, Science and Technology (MEXT), 3113 schools closed between 2004 and

2014, and about 300 schools on average close per year. There were 355 public elementary and junior high schools closed nationwide in fiscal year 2014; at 27 schools, Hokkaido had the most significant drop.

The establishment and/or consolidation of elementary and junior high schools are determined for each municipality based on the number of pupils in a given class and the number of classes in a given grade. MEXT has issued administrative guidance on the consolidation of government schools twice in the past. The first was in 1956 and encouraged the integration of schools. However, in a reversal in 1973, the agency indicated that “unreasonable school integration emphasizing scale was not to be carried out” and “small schools have an educational advantage, so it is preferable to leave schools small and enhance them.”[18] Even when consolidating, the ministry required: the maintenance of measures for students; a safe commute and keeping time for activities because schools play an important role in the community; and local residents’ agreement is necessary. As previously mentioned, schools are used not only for education but also for social activities, sports, as a meeting place, and festival sites for communities.

Between 2005 and 2010, the fluctuation in the number of elementary and junior high schools and the 5- to 14-year-old population by prefecture has been relatively consistent. Both Aomori and Akita Prefectures in the northeast have a high closure rate of over 10%. The impact consolidating schools has on communities is significant. When elementary and junior high schools are merged, the school intake range broadens, making commuting to school an issue for students. Operating school buses places a further financial burden on the municipality. Additionally, it is possible that families with children will leave the area if there are no longer sufficient educational opportunities. Moreover, there is concern that community vibrancy will be lost since it reduces overall opportunities for residents to gather and work together, such as sweeping streets and public parks for sanitation, and holding local cultural activities and festivals.

3.4.5.2 Senior High Schools, Universities, and Vocational Schools

In the midst of this drop in the youth population, young people are leaving to enroll in colleges, universities, and even senior high school in urban areas, creating a further significant decrease in enrolments at local higher education institutions. Colleges and universities play a major role in regional areas in ensuring a young population. If they withdraw, a further exodus of young people should be expected. Unique and attractive educational services are essential for the maintenance and survival of local higher education institutions. For example, Akita International University in Akita Prefecture was established in 2004 with the principle of boosting international communication skills through foreign language education. Students now come from all over Japan due to its fine reputation as a complete language education program, which includes lectures in English. The initial admissions quota of 100 students per year has gradually been increased to 175 today.[19] Another example is Hokkaido's Mikasa High School, which is the only high school in Mikasa City, whose survival was temporarily threatened by a rapid population decline due to the closure of the local coalmines. However, the school transformed itself into a specialist institute by introducing cookery and confectionery courses, as well as by building a dormitory. The school has now become the pride of the city, winning a national cooking contest among other awards.[20]

Finally, the village of Otoineppu, Hokkaido's smallest municipality, is the home of renowned craftsman Sunazawa Bikky, and the town established a municipal arts and crafts high school. Every year it draws in students even from outside Hokkaido, including 16 of the 118 students in fiscal year 2013. These are a few examples but they demonstrate the ingenuity necessary to attract new college and high school students, and these educational sites are contributing to maintaining a young populace in these areas. However, the number of young people who stay behind after graduation is low so the challenge is how to get them to stay.

3.4.6 Community Buses

In sparsely populated areas, operating public transport with a reduced number of users is increasingly problematic, and the cessation of bus and railway routes as well as a reduction in services is an impediment. As a result, there are already many areas where the daily life of residents, such as traveling to school and hospitals, has been hindered.[21]

Public transport not only ensures travel but communication between residents across villages and regions, it also provides necessary social cohesion that prevents isolation. In areas with no means of transport, community buses are being introduced as a countermeasure. The community bus introduced in 1995 in Musashino City, Tokyo, was one such catalyst.[22] While there is no clear definition of community bussing, they run "buses operated by the municipality itself and aimed at town revitalization by eliminating traffic vacuum zones and inconvenient areas, designed to improve resident welfare, and to promote outings by the elderly and the use of public facilities." With the amendment of the Road Transport Law in 2006 on-demand bus pools came into being.[23] Although conflicts with existing bus route remain an issue in urban areas, the problem in sparsely populated areas is how to increase the number of users overall. In many areas, proactive discussions have taken place between users and destination facility managers, leading to innovations where not only buses but all types of vehicles, including taxis, are used. The following cases are three comparatively successful municipal examples.

The community bus in Shizukuishi town, Iwate Prefecture (population approximately 17,500), is a demand-type taxi pool employing normal taxis, with a non-profit organization (NPO) commissioned by the town to manage it, which in turn commissions the local taxi company to provide the service. The six bus lines and eight routes running through the town were all abolished in 2004, after extensive consultation with residents, medical institutions, educators, senior clubs, and other groups in the previous year.[24] Six routes using a nine-seat jumbo taxi and small taxis were redrawn, reservations being made by taxi radio. After listening to users' opinions when reviewing the operational setup, a free direct phone link to the reservation center was established at the town clinic and the municipal offices. Routes have now been extended and the fare established at 200 yen or about 2 US dollars. It should be noted that the town managed to resolve this problem into a win-win relationship without competing with the town taxi company.

In the town of Tobetsu, Hokkaido (population of about 17,200), the town school bus and hospital, company and university shuttle buses all ran their own routes catering to specific users. These were integrated into a jointly operated community bus that the general population could also ride. Test runs began in 2006, and it is now firmly established as a mode of transport for residents. In a first for Japan, this public–private cooperated service, Tobetsu Commune Bus, has improved the finances of the bus operator. Fares are also set at 200 yen per ride, with free vouchers being provided to users of the former free hospital shuttle bus, thereby preventing a drop in services. This is also a successful example of pooling the local social resources of a region.[25]

Maizuru City, in Kyoto Prefecture (population of some 84,300), has its local resident organizations (councils) operate the community buses. In response to the private bus company's intention to withdraw from unprofitable routes, seven residents' councils established themselves as legal agencies to operate them. Each council decided on the most convenient ones and timetables for traveling to schools and hospitals, operating between two and four services a day. Meanwhile, areas that the councils deemed they did not need to pay for remain without transportation. This is an example of residents' councils acquiring corporate status and operating buses and is a useful future reference for small regions with few companies in place.

Public transport closure will not only take place in mountainous and depopulated areas, but may also occur in urban areas where a population decline means a drop in the number of users.[26] Although community buses have spread all over the country, not all transport vacant zones have been eliminated. There are still many areas where bus services continue to be financially supported by local governments. Many residents would likely agree that the financial expenditure to ensure their means of transport is worth it. At the same time, however, constant consultation with residents and innovative practices are required. Each region will need to make their own efforts using cases from across the nation as a point of reference.

3.5 Using IT as a Means of Escape

Based on the earlier assessments, the future for Japan looks bleak: the supermarkets are closed, the schools are gone (or are far away), and the buses do not run. What will life look like in areas like this? Both positive and negative outlooks are possible.

In the optimistic scenario, the Internet will allow people to survive without going to shops. People who will be 65 in 2040, the generation following the baby boomers, are currently 40 years old. Their information literacy is high and probably many will also be accustomed to shopping online. With a convenience store still near their home, they can buy their few daily or perishable provisions, drop off and pick up parcels, and even make withdrawals and payments at the ATM. Because schools are small, teachers and students will maintain a close relationship. Digital links with urban schools through the Internet will make it possible for young students to receive an interactive education. A few times a year, children from rural and urban areas will still be able to gather at a single location for classes so that their experience is not entirely virtual.

In the pessimistic scenario, residents seeking amenities will steadily continue to leave. With inconvenient day-to-day shopping, a shortage of childcare services and difficult child-rearing conditions, the outward migration of this generation will increase. There will be no gas stations or mobile shops, and the home delivery business, subject to significant losses in efficiency due to depopulation, will no longer deliver or pick up packages. The time will come when local residents will have to decide what sort of region they want. However, it may be too late to rectify the situation if young people have already moved away in droves. Even if steps are taken immediately, consensus building between residents and businesses takes time, and the actual results will only emerge years in the future. This is precisely why urgent measures are needed *now* with an eye to future population decline.

3.6 Single Seniors and Isolation

Variations in Japan's demographic structure will also change the composition of families. The most prominent change will be an increase in single households—the "singleization" of society. The number of single-person domiciles has been growing for some time in Japan, due to a changing family consciousness and a shifting socioeconomic environment. This trend will advance further in the future. According to the *Future Projection of Household Numbers in Japan* released by the IPSS, the total number of households in Japan will continue to increase until 2019, peaking at about 53 million. It will then start to decline, falling below 50 million by 2035.[27] The fact that the number of households will increase, while the population falls, means that the number of people per household will drop. The average number of household members is predicted to decline from 2.42 in 2010 to 2.20 in 2035. In addition, the number of single-person households will continue to increase from 16.8 million in 2010 to a peak at about 18.7 million in 2030, and thereafter begin to decline. The proportion of one-person households in general households will rise 4.8 points from 32.4% in 2010 to 37.2% in 2035 (Fig. 3.12).[28]

In the future, both rural and urban areas will experience an increase in the number of elderly living alone. The nationwide share of one-person households that are "elderly households" was already 30.7% in 2010. This is expected to advance in both rural and urban areas by 2035 in Tokyo (44.0%), Osaka (43.8%), Kagoshima (43.1%), and Kochi

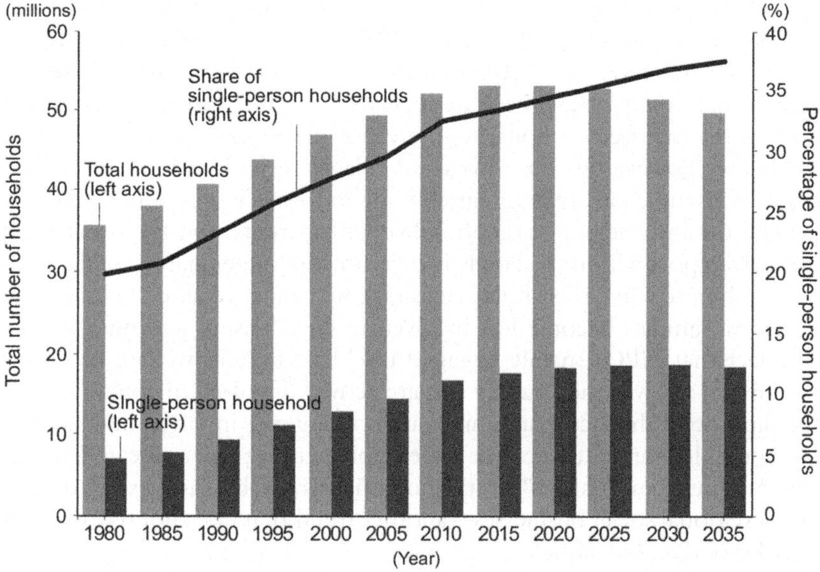

Fig. 3.12 Number of households and the percentage of single-person households (estimate). Source: IPSS, Future Projections of Household for Japan (National Estimate), January 2013

Prefecture (42.7%). The cause of this increase is that elderly couples, whose children have moved to urban areas, become single elder households after the inevitable death of one spouse. Young people that have remained in the region without migrating to urban areas also become one-person households because they often lack marriage opportunities. According to the *2011 Survey on Marriage and Family*, conducted by the Cabinet Office, 39.3% of single males in their twenties and 31.2% of single males in their thirties living in regional areas "have never associated with the opposite sex."[29] With marriage coming later or never in life, young people remain single longer and continue to live alone. A significant portion of these young people will age and may become single seniors. This increase in single households will be a factor in the isolation of residents, raising the danger of an "unconnected society." This problem is already surfacing in the form of "isolated deaths" of the elderly, especially in suburban housing complexes where community ties are already weak (Tachibana 2011).

Numerous young people are already isolated and withdrawn from society. Reports suggest that 113 out of 1293 inhabitants aged between 18 and 55 in the town of Fujisato, Akita Prefecture (population of some 3900), or approximately 10%, have become recluses.[30] Changes in family structure, such as the one-person household, are not necessarily taking place across the board. For example, in towns and villages in Hokkaido with relatively larger households as units of agricultural and fishery production, the next generation and their spouses often live with parents. On the other hand, mutual support within the family is currently no longer viable. As the norm for children to leave their homes and take up residence alone in large cities develops, families become less involved in child raising and nursing care. The fact that NPOs mainly engaged in elder care and welfare have been established all over the country and are actively involved in communities is the flip side of the fact that families are no longer as involved in child-rearing and elder care. At the same time implementing elder care and welfare based on new "mutual aid" has the potential, as a viable business, to develop into a regional social businesses—another trend that does not rely on traditional blood and local ties.

In this chapter, we have tried to portray how reductions in infrastructure, particularly in rural areas and caused by population decline, will impact on the daily lives of Japanese throughout the archipelago. Depopulation shrinks regional markets and reduces employment. This then triggers further population outflows and spurs a collapse in the infrastructure that undergirds a dynamic society. Falling into this negative chain reaction will fulfill the "pessimistic scenario" we outlined before. However, with foresight and planning, the possibility of addressing these community concerns through directed NPOs and new businesses have also emerged in certain areas of Japan as a response. This approach perceives these difficult issues as potential opportunities, and could be seen as a more "optimistic scenario" for the future.

Both the optimistic and pessimistic choices are possible but it is up to each and every resident to decide which alternative to choose.

Notes

1. MLIT 2014.
2. IPSS 2013b.
3. Made own projection beyond 2040 by extrapolation using the same assumptions, analyzed by Igarashi, Chikako, President, HIT (IPSS 2013b).

4. MIC 2009.
5. MHLW (2014).
6. MLIT 2014; MIC 2015.
7. AIRIA 2015.
8. Analysis using *2040 Micro Population Data* and *2010 Phone Directory DB Telepoints with Coordinates* (provided by the University of Tokyo, CSIS, hereinafter referred to as "digital phone directory").
9. According to the digital phone directory.
10. [Method of analysis] Micro population data uses mesh data to allocate the population at a building level using 2040 population projections and map information, the population estimates being performed by the Hokkaido Intellect Tank (HIT) and the map information data analysis by the University of Tokyo's Center for Spatial Information Science (CSIS). Figure 3.13 is a flow chart on how the data was created. It should also be noted that because of the difficulty in predicting building distribution in 2040, it was assumed that the distribution and usage of buildings in 2010 would pertain (NB: The survey period of house mapping was 2008 to 2009).

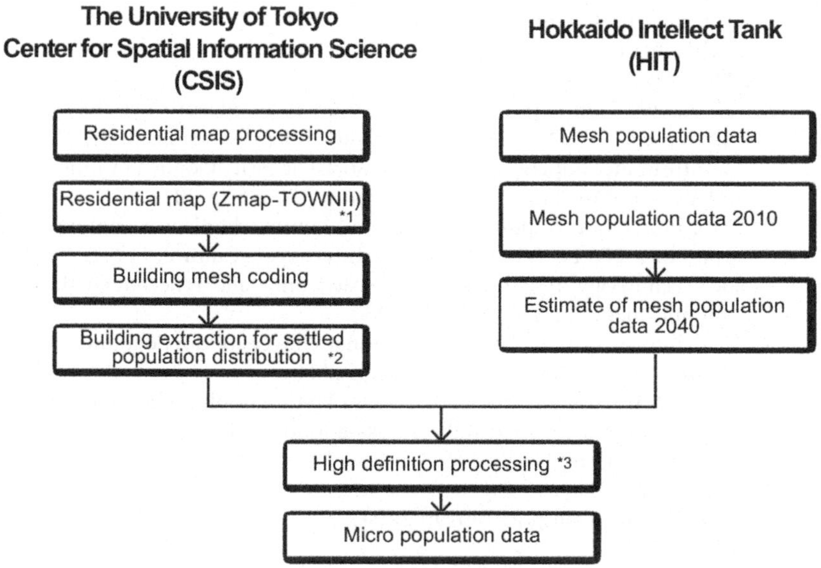

Fig. 3.13 Flow chart of the method of analysis

(*1) Residential map building polygons carry attribute information, such as office, residential, shared building.
(*2) Buildings for settled population distribution are extracted. They are divided into residential, shared building or other. If they are none of these, they become candidate allocation destinations for businesses.
(*3) The total floor area of a building is determined by its floor space and the number of floors, with settled population allocated thereto.
(Note) The above technique is an application of part of the non-aggregate reduction and allocation method for existing demographics.

11. In fact, a detailed trading area for each store has been set and in many cases, even in small municipalities, just because you are out of the 5-kilometer radius does not necessarily mean you are outside the trading area.
12. Estimated from the microlevel population projection data produced by the CSIS, University of Tokyo, by combining the 2010 microlevel population data produced by Akiyama Yuki, assistant professor at CSIS, University of Tokyo, and the 500 m mesh aggregated population projection data (2010–2040) produced by HIT; general incorporated association and the 2010 Digital Phonebook (number of supermarket stores).
13. JFA 2015.
14. Estimated data produced in combination from CSIS, University of Tokyo, HIT and 2010 Digital Phonebook.
15. ANRE 2014.
16. MIC 2008.
17. "Integrated community care" had existed as a concept since the bottom half of the 1970s. The system itself came under discussions in the Nursing Care Insurance Subcommittee of the Social Security Council in June 2010, and an aspirational provision was added to the law in 2011.
18. Since integration of public primary and junior schools was announced in September 27, 1973 by Minister of Education, Culture, Sports, Science and Technology, the concept was passed onto and reflected in the new announcement in 2015 (MEXT 2015).
19. More information on Akita International University can be found on their website: http://web.aiu.ac.jp/undergraduate
20. More information on Hokkaido Mikasa High school can be found in their website: http://www.city.mikasa.hokkaido.jp/highschool/
21. Otoineppu 2015.
22. MLIT n.d.
23. Kinki District Transport Bureau 2004.
24. MLIT 2007.
25. Tobetsu n.d.
26. MLIT 2012.
27. IPSS 2013a.

28. IPSS 2013a.
29. IPSS 2011.
30. Cabinet Office 2010.

References

Automobile Inspection & Registration Information Association. (2015). 1.069 Automobiles per household—Ratio stays flat from previous year (Issetai atari 1.069 dai de zennendosuijun) (Press release, August). http://www.airia.or.jp/publish/file/r5c6pv0000003pun-att/r5c6pv0000003pv2.pdf, accessed on September 25, 2016 (in Japanese).

ANRE. (2014, July 8). Trends of volatile oil distributors and the number of gas stations based on registration (Kihatsuyu hanbaigyosha oyobi kyuyujo no suii). http://www.enecho.meti.go.jp/category/resources_and_fuel/distribution/hinnkakuhou/data/20140708_ss25fy.pdf, accessed on September 25, 2016 (in Japanese).

Cabinet Office, Government of Japan. (2010, July). Research of Social withdrawal (Wakamono ishikini kansuru chosa (hikikomori ni kansuru jittai chosa) hokokusho). http://www8.cao.go.jp/youth/kenkyu/hikikomori/pdf/gaiyo.pdf, accessed on September 25, 2016 (in Japanese).

Cabinet Office, Government of Japan. (2014, December). Headquarters for Overcoming Decline and Vitalizing Local Economy in Japan, Japan's Long-term Vision and Comprehensive Strategy, http://www.kantei.go.jp/jp/singi/sousei/info/pdf/panf_eng.pdf, accessed on February 16, 2017.

IPSS. (2011). Attitudes toward marriage and family among Japanese singles: The fourteenth Japanese national fertility survey in 2010—highlights of the survey results on singles. http://www.ipss.go.jp/site-ad/index_english/nfs14/Nfs14_Singles_Eng.pdf, accessed on September 25, 2016.

IPSS. (2012). Japan's population forecast (Nihon no shorai suikei jinko), January 2012. http://www.ipss.go.jp/syoushika/tohkei/newest04/sh2401top.html, accessed on August 14, 2017.

IPSS. (2013a). Future Projections of Household for Japan (National Estimate) (Nihon no setai su no shorai suikei zenkoku suikei), January 2013. http://www.ipss.go.jp/pp-ajsetai/j/HPRJ2013/t-page.asp, accessed on August 14, 2017

IPSS. (2013b). Japan's population forecast by region (Nihon no chiiki-betsu shorai suikei jinko). http://www.ipss.go.jp/pp-shicyoson/j/shicyoson13/1kouhyo/gaiyo.pdf, accessed on 2016, September 22 (in Japanese).

JFA. (2015). Convenience store statistics data (Konbiniensu sutoa tokei deta). http://www.jfa-fc.or.jp/particle/320.html, accessed on September 25, 2016 (in Japanese).

Kinki District Transport Bureau. (2004, March). Community Bus Introduction Guide.

MEXT. (2015, January 27). Notice by Minister of education, culture, sports, science and technology. http://www.mext.go.jp/a_menu/shotou/shugaku/detail/1354768.htm, accessed on September 25, 2016 (in Japanese).
MHLW (2014). Vital Statistics of Japan. (2013-nen Jinkou Doutai Hakusyo kakutei chi no gaikyou, Hyou 3-2) http://www.mhlw.go.jp/toukei/saikin/hw/jinkou/kakutei13/index.html, accessed on August 8, 2017
MIC. (2009). Statistics Bureau, 2009 Economic census for business frame. http://www.stat.go.jp/english/data/e-census/, accessed on September 22, 2016.
MIC. (2008). Excellent business cases of year 2008 (Heisei 20-nendo yuryo jirei shu). http://www.soumu.go.jp/main_content/000063250.pdf, accessed on September 25, 2016 (in Japanese).
MIC. (2014). Statistics Bureau, Report on Internal Migration in Japan 2014, http://www.e-stat.go.jp/SG1/estat/ListE.do?lid=000001129143, accessed on February 16, 2017.
MIC. (2015). Future with ICT (Heisei 27nen hakusho, dainibu, ICT ga hiraku mirai). http://www.soumu.go.jp/johotsusintokei/whitepaper/ja/h27/html/nc231110.html, accessed on September 25, 2016 (in Japanese).
MLIT. (2007). On-demand bus service using taxi (Ippan no takusi wo riyoshita demando sabisu). http://www.mlit.go.jp/sogoseisaku/transport/pdf/010_shizukuishi.pdf, accessed on September 25, 2016 (in Japanese).
MLIT. (2012). Manual to sustain, maintain and improve regional public transportation (Chiiki kokyo kotsu kakuho iji kaizen ni muketa torikumi manyuaru). http://wwwtb.mlit.go.jp/kinki/kansai/program/manual.htm, accessed on September 25, 2016 (in Japanese).
MLIT. (2014). Grand design of national spatial development towards 2050, Japan—Creation of a country generating diverse synergies among regions (Kokudo no gurando dezain 2050). http://www.mlit.go.jp/common/001088248.pdf, accessed on September 22, 2016.
MLIT. (n.d.). Community bus (Komyuniti basu). http://www.mlit.go.jp/jidosha/anzen/01transit/communitybus.html, accessed on September 25, 2016 (in Japanese).
Otoineppu Village. (2015, March). Comprehensive plan of Otoineppu village, No.5 (Otoineppu mura sogo keikaku dai 5), http://www.vill.otoineppu.hokkaido.jp/gyousei/keikaku/files/otoineppu_general_plan.pdf, accessed on September 25, 2016 (in Japanese).
Tachibana, Toshiaki. (2011). *Isolated Society—how networking of family, community and work has collapsed (Muenshakai no shotai ketsuen chien shaen wa ikanihokaishitaka)*. Tokyo: PHP.
Tobetsu. (n.d.). Tobetsu community bus (Tobetsu fureai basu). https://www.town.tobetsu.hokkaido.jp/site/fureai-bus/, accessed on September 25, 2016.

CHAPTER 4

The Impact of Demographic Changes on Macroeconomic and Public Finance

Oguro Kazumasa

When making predictions about Japan's economic growth and considering growth strategy initiatives, one of the most important structural changes to keep in mind is the change in the demographic structure due to a declining birth rate and aging population. In Japan, the working-age population (people between 15 and 64 years of age) peaked in 1995, and it has fallen consistently since then. The total population also peaked in 2008, and subsequently began to decline. The rate of aging that indicates the ratio of those aged 65 or more in the total population, which was about 12% in 1990, had reached approximately 23% in 2010. This aging and declining population structure is expected to continue over the next few decades, and is developing into one of the most fundamental constraints on Japan's future social and economic sectors.[1]

4.1 The Key Is "Real Growth Rate per Capita"

The impact on the economy of the population composition getting older can be considered separately in terms of production and consumption. Looking first at the production side of the equation, due to a decrease in Japan's working-age population, in order to keep the real gross domestic

Oguro. K (✉)
Faculty of Economics, Hosei University, Tokyo, Japan

Y. Funabashi (ed.), *Japan's Population Implosion*,
https://doi.org/10.1007/978-981-10-4983-5_4

product (GDP) growth rate positive, labor productivity (value-added production per worker) must increase greater than the decrease in the labor force because it is this force itself that creates added value. For example, although the real growth rate during the 1980s averaged 4.7%, this fell to an average of 1.1% in the 1990s, and remained sluggish thereafter at around 0.7% (Oguro 2014). In addition, the government's November 2014 Council on Economic and Fiscal Policy, Expert Panel, Committee for Choice for the Future released a final report where the economic outlook in 50 years projected low growth of −0.1% on average annually from 2040 if depopulation continued at similar rates and productivity remained torpid. This was further confirmed by an article in the *Nihon Keizai Shimbun.*[2]

Given that it would be difficult for Japan to bolster its national real GDP growth rate in such a situation, and the fact that some speculate that only luck can even maintain the GDP level per capita, it is necessary to acknowledge the grim prospects that a falling working-age population will have on economic growth (Fig. 4.1).

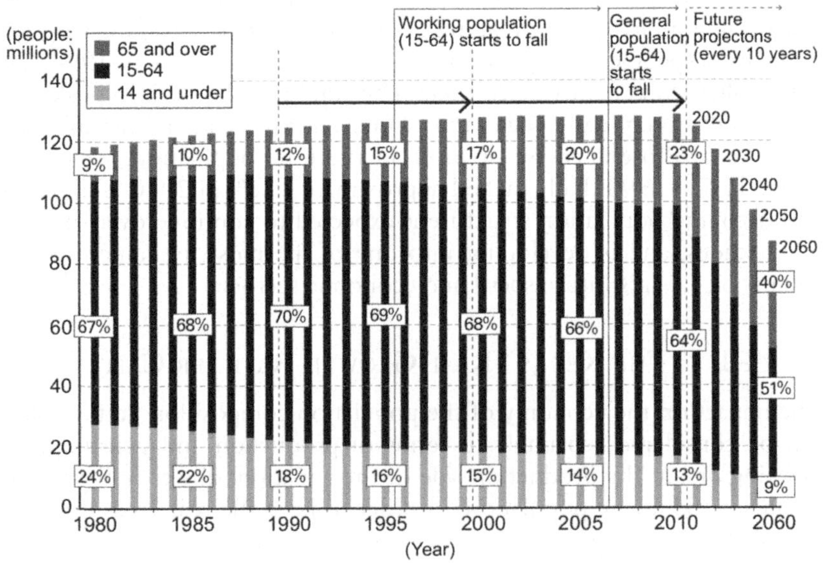

Fig. 4.1 Changes in Japan's demographic structure. Source: MIC, National Census; Intercensal Adjustment for Current Population Estimates; Population Projection for Japan (January 2012)

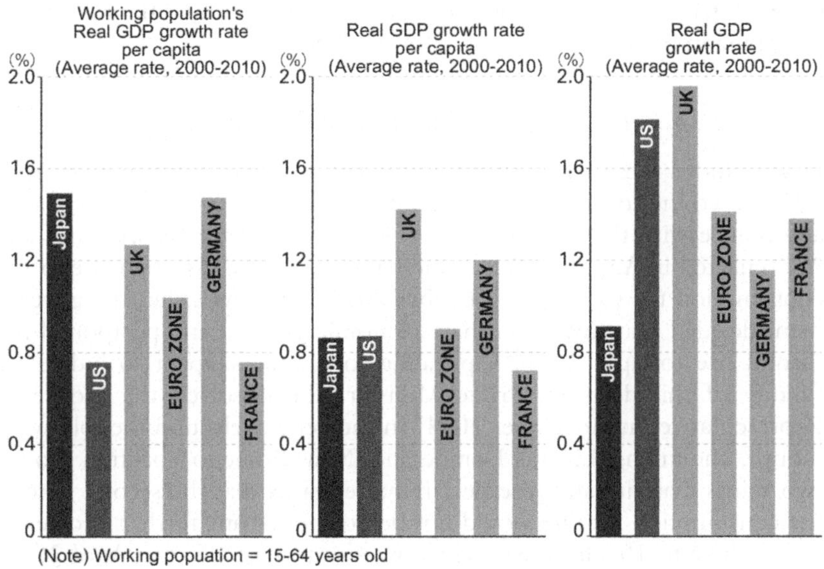

Fig. 4.2 International comparison of real GDP growth rates. Source: World Bank, Haver Analytics

The Japanese economy is currently beset by a sense of distress. The real growth rate continues to fall and Japan had the lowest figures compared to the annual averages in the USA, the UK, the Eurozone, Germany, and France for 2000–2010 (Fig. 4.2). Estimates show Japan's real growth rate turning negative from 2020–2040.[3] However, there is no need to be overly pessimistic. What is more important than real growth rates is the tabulation for "real growth rate per capita," which provides an indication of the improvement in living standards. Japan's real growth rate per capita was roughly the same rate as that of the USA between 2000 and 2010, and Japan's real growth rate per capita of the working-age population, representing the growth of per worker productivity, was also the highest compared with other Organisation for Economic Co-operation and Development (OECD) nations (Fig. 4.2). Japan's average real GDP growth rate per capita (annual rate) for 2003–2012 was 0.82%. Looking at other countries, it was 0.96% in the USA, 0.45% in France, 1.31% in Germany, –0.67% in Italy, 0.64% in the UK, and 1.4% in Australia. This period includes the impact of the Great East Japan Earthquake

(March 2011), and excluding that year, the average real GDP growth rate per capita in both Japan and the USA was 0.94%.

4.2 The "Abenomics" Fantasy

It was under these circumstances that the Abe administration in April 2014 raised the consumption tax rate from 5% to 8%. This measure was in accordance with the Comprehensive Reform of Social Security and Tax Act (enacted in August 2012), which also scheduled a second rise in the consumption tax to 10% for October 2015.[4] However, the Abe government decided that this would halt economic growth, and postponed the second rise to April 2017.[5] Appealing for public support to judge the validity of this decision, Prime Minister Abe dissolved the House of Representatives in November 2014. In the general election the following month, the ruling Liberal Democratic/New Komeito coalition won a two-thirds majority and Abe declared afterwards that the second hike in the consumption tax rate would not be postponed any further. However, the tax hike to 10% from 8%, which was set to take effect in April 2017, was postponed to October 2019. In addition, just because it will be raised does not mean that the fiscal deficit will be eliminated. The increase in fiscal resources obtained from a 5% rate to a 10% rate is approximately 13.5 trillion yen. On the other hand, social security benefit costs are currently expanding at a pace of 2.6 trillion yen a year. With the revenue of social insurance premiums remaining flat and benefit costs continuing to expand, the entire amount of the increased portion will be borne by public expense, completely devouring the increased tax revenue in about five years. Revamping public finances inevitably will mean the additional "pain" of cuts in social security costs and higher taxes.

The problem of interest payments, which will manifest in the future, is another factor that will exacerbate the budget deficit. Although long-term interest rates have been reduced through the Bank of Japan's monetary easing, it is predicted that even if the level of interest rates stays stable, the surge in public debt will lead to an expansion in interest payments from the current 9 trillion yen to approximately 17 trillion yen per year by fiscal 2019.[6] Attempting to stabilize public finances without limiting social security costs would require the consumption tax to be raised much higher than 10%. It is helpful in this context to consider the estimates of Richard Braun of the US Federal Reserve Bank of Atlanta and Professor Douglas Joines of the University of Southern California (Braun and Joines 2014). According to their calculations, should the Abe administration fail to realize its 2% inflation target and try to stabilize public finances with a sharp increase to the consumption tax rate in 2017, the tax rate would ultimately need to be raised to 33%. In addition, if the consumption tax increase was implemented in 2022 rather than 2017, the final tax rate is estimated to reach 37.5%. To maintain healthy public finances, the tax rate needs to be raised by approximately 1% for every year that the tax increase is postponed. This can be seen as the cost of putting off reform.

Unless inflation can be sustained at 2%, Professor Gary Hansen and his colleagues at the University of California, Los Angeles, also estimate that the final tax rate required for financial stabilization would be 35% (Hansen and Imrohoroglu 2016). In other research, Professor Oguro Kazumasa of Hosei University and Shimasawa Manabu, a team leader at the Chubu Region Institute for Social and Economic Research, also estimated the required tax rate needed to be 33.5% at its peak to cover deficits (Oguro and Shimasawa 2011).

What will happen if Abenomics were successful and the 2% inflation target is actually achieved? Fiscal drag, for example, would still exist even if real income remained unchanged but nominal income would increase under the progressive income tax system to create a real tax increase.[7] As a result, moderate inflation would certainly lead to indirect public support for a monetary policy that included fiscal and social security reform, but it would still be necessary to raise the consumption tax rate. The research of Braun and his colleagues posits that even in such a case, the final consumption tax would have to be set at 25.5% if a single tax hike were to be implemented in 2017. In practice, however, it is more difficult to raise the consumption tax in one steep incline than to raise it in intervals. Therefore, Braun and his team also examine a scenario where: (1) 2% inflation is realized; (2) co-payments for the elderly are increased to 20%; (3) the guarantee of pension payments equivalent to half an annual income when working is eliminated; (4) government operating expenses are cut by 1% as a package and the consumption tax rate is raised in stages, and then lowered after the aging peak has passed. As a result, the consumption tax rate would reach 32% at the peak (Fig. 4.3).

This consumption tax rate does not take reduced tax rates for food and other necessities into account. If a reduced rate were introduced, an even higher rate would be required to compensate for lost revenue. Moreover, a gradual reduction in the effective corporate tax rate, from 34.62% to 29.74% by fiscal 2018, was enacted in fiscal 2015. If an alternative source of revenue for this tax cut cannot be secured, the consumption tax rate will probably have to be higher. This means that regardless of whether the

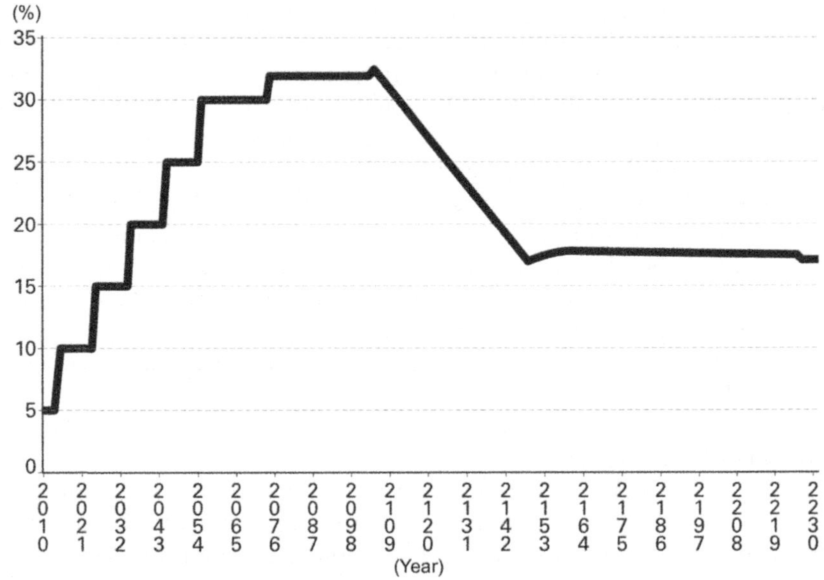

Fig. 4.3 Simulation result for the case of a gradual hike in consumption tax rate. Source: Revised from Braun and Joines (2011)

inflation target is achieved or not, the final consumption tax rate will have to be in excess of 32%. It is obvious from the above calculations that the more Japan puts off drastic fiscal and social security reform, the greater the cost of any subsequent reform, which makes the position that Japan's economy will make a painless comeback if 2% inflation can be achieved untenable. The Japanese economy is currently more positively framed thanks in part to rising stock prices and Tokyo's selection for the 2020 Olympic Games. However, in order to avoid a financial crisis, the pain of additional cuts in social security and further tax increases cannot be avoided. This is the reality.

4.3 Fiscal Crisis Accelerated by a Sharp Increase in Social Security Benefits

Japan's public debt now tops 200% of its GDP. This scale surpasses the level faced by the nation at the end of World War II, which led to rampant inflation following the surrender (Fig. 4.4). Immediately after the war, the

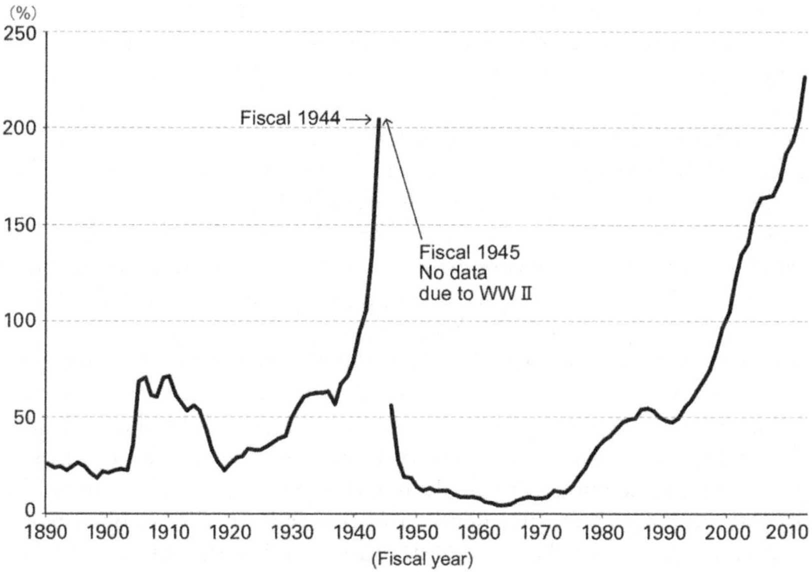

Fig. 4.4 Percent change in Japan's debt to GDP ratio. Source: Statistics Department, the Bank of Japan (for figures before 1946); Hundred Year Statistics of the Japanese Economy, Annual reports and others (for figures after 1947)

government imposed a "special tax on wartime indemnity" to pay recipients of wartime compensation the same amount as their past payment (Takahashi 2012).[8] As inflation quickly intensified, the government introduced the "conversion to a new yen" plan, shut down bank deposits, and conducted surveys on property for confiscation. There is an increasing probability of financial collapse when government debt exceeds 200%, although it does not necessarily mean that a collapse of public finances is imminent. The costs of social security benefits are a central factor in this crisis. Since the acceleration of aging is faster than Japan's current economic growth rate, benefit costs continue to outpace everything else. Social security benefits, which stood at approximately 84 trillion yen in fiscal 2003, swelled to approximately 110 trillion yen in the 10 years up to fiscal 2013 due to aging. This amounts to more than 20% of Japan's nominal GDP. Although the speed at which social security benefits expand fluctuates from year to year, the annual average increase has been 2.6 trillion yen over the past decade. There is a common idea in Japanese society that an increase of about 1 trillion yen a year in the costs of new social security benefits is normal, but this is a far too optimistic view. Funding for these benefits as seen in fiscal 2013—approximately 110 trillion yen in total—consisted of about 60 trillion yen in social insurance premiums and approximately 10 trillion yen in asset management revenue; the approximately 40 trillion yen or so remaining came from the public purse. About 1 trillion yen of this public expense from growth in "social security-related expenses" was borne by the state; from this assessment the media has mainly created the commonly held notions about future costs. If, on the one hand, the cost of benefits continues to expand by 2.6 trillion yen a year, and, on the other hand, social insurance premiums remain flat, the possibility of social security-related costs also increasing to the same extent as benefit costs can no longer be ignored. Moreover, since the public burden cannot be covered by tax revenue, the majority of the shortfall will be passed on to future generations through the issuance of government bonds. The same applies for medical and nursing care. For example, 50% of the basic pension payments are covered at the public expense (supplemented from the general account), and if the amount of social security benefits surges as demographic aging continues, public expense will thus automatically surge as well.

Additionally, the increase in the proportion of the elderly in the population may have a negative impact on the economy by reducing household savings. Japan's abundant financial assets held by households, 1741 trillion yen, have been deposited in banks and serve as the funds that, as the source of bank loans to businesses, have been the foundation of Japan's economic growth.[9] However, given that there is a strong general trend for the elderly to draw down their savings rather than keep these funds in the banks, a low birth rate and demographic aging have the effect of lowering the household savings rate as a whole. Should the birth rate continue to fall while, proportionately, society continues to age, the household savings rate will shift into the negative and there will be a drop in household financial assets. This risk will dampen corporate investment and have a negative impact on Japan's economic growth. However, it is likely that the slowdown in domestic corporate investment can be attributed more to the impact of depopulation on domestic consumption rather than the influence of a drop in household savings. Moreover, even if household savings decrease, leaving insufficient funds to absorb corporate and government bonds domestically, it is logically possible to look to foreign savings for that purpose.

Apart from an aging demographic structure increasing the burden on the working generation through higher social insurance premiums and tax burdens to support social security, in terms of consumption, the potential for growing anxiety among the labor force about the sustainability and future growth potential of the social security system will probably lower consumption if the government tries to maintain the existing level of social security. Moreover, there naturally will be a limit to any expansion in domestic demand given the fact that Japan's population is shrinking. A general fear that this fall in consumption will trigger a negative chain reaction along the lines of lower corporate growth expectations and cause a drop in investment, thus further dampening the economy, according to the National Institute of Population and Social Security Research.[10] In this way, although an aging and declining population can serve as major constraints on Japanese economic growth, as long as the pertinent causes of the trend for falling birth rates and an aging demographic (or, more precisely, greater longevity) remain to some extent irreversible, we will need to think about how to overcome these problems and not just descend into blanket pessimism.

4.4 Responding to Increasingly Important Emerging Markets

Another significant structural change confronting Japan's economy is the rapid growth of Asia and emerging markets. This rise has several implications for the Japanese economy. The first implication, especially for the manufacturing industry, is that the rise of companies in these new markets will make competitive conditions tougher for Japanese firms, and could lead to a further decline in profitability. For example, the South Korean company Samsung was in first place in the flat-screen TV global market with an overwhelming share of 23.8% in 2011 and was followed by another South Korean firm, LG with 13.7%. On the other hand, Japanese firms such as Sony (10.6%), Panasonic (7.8%), and Sharp (6.8%) only ranked in the third to fifth positions. This situation in the flat-screen TV market was also reflected in overall corporate performance. The recording balance sheet deficits of the three Japanese firms of Sony, Panasonic, and Sharp show a loss of several hundred billion yen in the fiscal year ending March 2012. This news sent shock waves throughout the Japanese economy. At the same time, one positive impact of the growth of emerging economies on the Japanese economy is the concurrent rapid expansion from the

Asian markets in particular. Expanding markets in the Asian region with their close geographical proximity to Japan may well be a great opportunity for realizing growth in Japan where there is a limit to any expansion of domestic demand as aging and depopulation progress. This potential applies not only to the manufacturing industry, but also to Japanese non-manufacturing industries, such as life insurance and other companies, convenience stores, and retail shops.

4.5 The Problem Is Low Labor Productivity

Another key point that should be mentioned when considering the growth of the Japanese economy is low labor productivity. The labor productivity growth rate in Japan plummeted from the early to late 1990s after the collapse of the bubble economy, and it is not high by international standards, ranked 21 out of the 34 OECD countries in 2014. Basically, labor productivity is determined by two elements: the amount of capital stock per worker and total factor productivity (TFP), which indicates the degree of technological progress. But in Japan today when capital stock per capita is already in slight excess, the key to productivity growth lies in a higher TFP, in other words, technological progress and innovation. A key point when considering growth strategies is that creating sustained technological progress and technological innovation is essential for future growth in the Japanese economy. In order to improve overall Japanese productivity levels, it is important to shift production from inefficient to efficient sectors, to improve the environment for a higher metabolic industrial structure, to foster a supportive environment for accelerating structural transformation, and to speed up the restructuring of industries. For example, greater corporate management transparency and corporate governance reform to raise return on equity are required to reform the labor market to promote greater fluidity. While innovation is a key issue, the question of industrial metabolism is even greater in Japan today and this is closely related to demographic issues and the regional problems it causes.

When viewed on an industry-by-industry basis, labor productivity in the service industry is lower than in other sectors. Since the services account for a growing share of the economy, their persistently low productivity is a significant problem for the Japanese economy (Fukao 2010). Additionally, in different areas of the service industry, sales per capita (labor productivity is represented by "value added per hour per person," so this is not exactly equivalent to sales per capita, but the two can generally be viewed as being correlated) are high in the entertainment and information communication industries, but attention needs to be paid to the fact that levels are low in health care and insurance/welfare, fields expected to grow due to falling birth rates and aging. The reason for low productivity in medical and nursing care appears to stem from the fact that the government conducts price controls through medical treatment fees, alongside its fiscal constraints. If this structure is maintained, productivity in the Japanese economy as a whole will drop even if the scale of the medical and nursing care market expands due to the demographics of declining birth rates and aging. There is a need, therefore, to raise productivity in these industries and turn them into a "profitable" sector commensurate with the market expansion. However, there will be increasing concern for fiscal collapse in the government sector, which could conversely hamper growth if payments from public health insurance and nursing care insurance are allowed to escalate out of hand accompanying such a move.

To link market expansion in this area to sustainable growth, it is important that entry regulations are eliminated as much as possible to encourage private sector participation and, at the same time that the range covered by the public purse is reviewed (Oguro 2015). In some areas it may become appropriate to ask users to pay themselves according to the services received and to consider removing those services from the public insurance list. For example, in the field of nursing, regulations could be relaxed and a private-sector-led market expansion promoted in place of removing advanced care equipment and nursing care robots, residence in nursing homes that exceed a certain level of floor space and amenities, and elder residential facilities that do not go quite as far as providing nursing care from the public long-term care insurance list. Regarding medical care, a major point of debate involves lifting the ban on so-called mixed medical care to provide a combination of care that is both covered and uncovered by health insurance. Under Japan's current system, mixed medical care is prohibited and if certain procedures not covered by insurance are provided, then the care will be treated in its entirety as uncovered (Ikegami 2006; Kawaguchi 2012).

Many pundits claim that the status quo prohibiting mixed medical care in fact creates greater inequity between the wealthy—enjoying medical care at their own expense—and those people with fewer resources. There is a strong need to examine the scope of public insurance coverage and to be more selective in concentrating limited medical resources by expanding mixed medical care and making certain areas of currently covered treatment the subject of "selective medical treatment," where patients can choose to pay an additional cost for given treatments, as well as reducing insurance coverage for generic over-the-counter prescriptions.[11] Along with using private medical insurance to supplement public insurance, this is an issue that needs to be examined instead of treating it as a taboo.

4.6 National Savings Trends to Be Watched

We should note that in relation to Japan's population dynamics, savings rates are experiencing a downward trend. This is also important and can be seen in the transition in "national savings" (relative to the Gross National Income (GNI) ratio) provided in Fig. 4.5. Here "national savings" represent total net savings excluding the portion of fixed capital consumption from total private and government savings, but national savings have been steadily losing ground gradually since 1990 after the collapse of the bubble economy. As demographic aging gained ground, it finally slipped into negative growth in 2009. It should be noted that "government savings" refers to the difference between general government

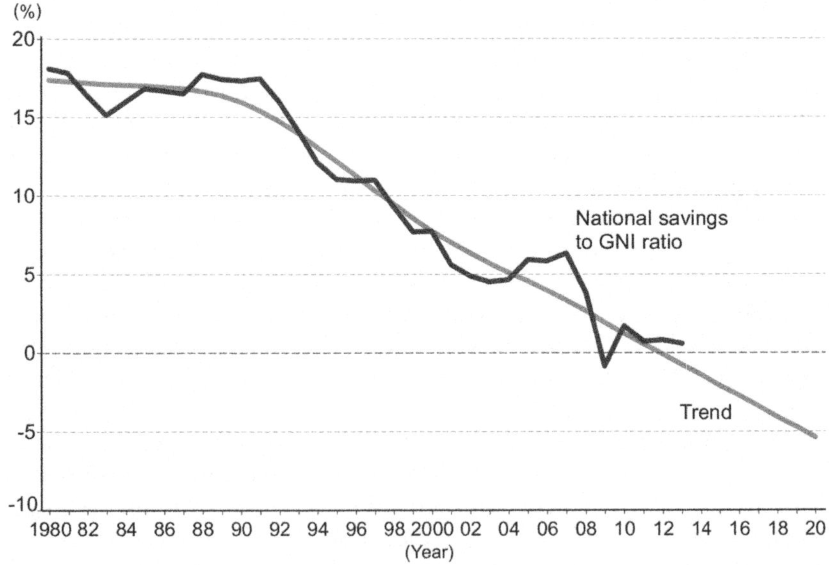

Fig. 4.5 Percent transition in Japan's national savings (relative to GNI ratio). Source: Cabinet Office, Annual Report on National Accounts, 2009 and 2013

revenue (national + local + social security funds)—such as tax and insurance premiums—and current government expenditure excluding investment in social infrastructure and the like, and since this is basically "government savings = the fiscal balance – government investment," the size of the government savings deficit corresponds to the volume of deficit bonds issued.

Based on the Investment Saving (IS) balance that represents the relationship between investment in a country's economy and savings, the IS balance equation—national savings = (investment – fixed capital consumption) + the current account—can be derived. However, since Japan's current account balance (which is still in the black at the moment although gradually shrinking) is in surplus, the fact that national savings, which amount to a transfer of wealth to the next generation, have fallen into the red means that the Japanese economy is currently in a situation where public debt issued to finance the budget deficit is eating away at private savings, and the private sector is no longer able to cover the cost of replenishing capital stock. Moreover, Japan's social security budget (pensions, medical and nursing care, etc.) incorporates an underlying mechanism that expands the deficit in government savings.[12] This is because the portion of social security benefits like pensions, health care, and long-term care that cannot be covered by insurance premiums are supplemented by public funds from the government's accounts, a figure which is expanding at a rate of about 1 trillion yen a year. This blowout in public fund compensation raises government expenditures, but if there are insufficient fiscal resources, it also increases the government savings deficit. The country's fiscal balance sheet is consistently in the red for government savings, but so too are those for the prefectures since 2002, and for social security funds since 1998. In short, the expansionary trend over the past few years is impossible to overlook. Should private savings stay flat or fall due to demographic aging, the national savings deficit obtained from the sum of private savings and government savings (excluding fixed capital consumption) will continue to expand. Such a drop in national savings means that running down capital stock would begin in earnest, which suggests an eventual scenario of shrinking domestic production.

4.7 Will a Government Bond Crash Occur?

Let's look briefly at a simulation of what would happen in the absence of fiscal and social security reform if long-term interest rates started to rise as demographic aging moves forward.

A rise in long-term interest rates means a drop in government bond prices. Once the decline starts, the market becomes trapped by fear from the uncertainty about whether it will fall further, creating a possibility for the social phenomenon of "cut-and-run." This effect, in turn, would see prices fall further and interest rates rise. While a rise in interest rates of about 1% could be somewhat absorbed, interest payments on debt would surge if the increase were any higher. For example, with a debt of 1000 trillion yen and inflation at 2%, interest payments would jump by 20 trillion yen by simple calculation should long-term interest rates go up by two percentage points. Strictly speaking, since interest payments on outstanding government bonds are already set, it would be interest payments on newly issued government bonds and refunding bonds (bonds issued to defer debt by raising funds through repayment on already issued government bonds) that would rise. In reality, payments would increase by 20 trillion yen over a period of several years. If long-term interest rates were to rise further, this would also increase interest payments commensurately. The heightened burden would undoubtedly do away lightly with consumption tax increases.

Realistically speaking, it would be difficult to cover an increase in interest payments of up to tens of trillions of yen through tax increases or spending cuts. This would be the "first phase" of the start of a fiscal crisis. Even with a few years' grace, should the market face the current fiscal situation directly, there is the possibility of a worst-case scenario where no buyers for government bonds emerge in the market. Or without going that far, government finances might no longer be able to stand the burden of interest payments with soaring long-term interest rates. At this point we would quickly enter the "second phase." Because the government could not afford to leave the situation as it stood, it would have to take emergency measures. One of them would be default. However, this is chosen in many cases when government bonds are held by overseas entities. It amounts to being prepared to let foreigners take a hit and to being exposed to condemnation from the international community in order to protect one's own people.

In Japan's case government bondholders are overwhelmingly domestic investors (financial institutions, such as banks), so there is a strong possibility that the domestic economy and financial system would descend into chaos if the government defaulted. Additionally, because Japan is involved in a deep and complex web of overseas monetary transactions in the global economy, the shock of a default would not be confined to Japan, but would inevitably have a significant impact on the global financial system. This is a choice Japan cannot make.

What about the direct underwriting of government bonds by the Bank of Japan? Article 5 of the Finance Act specifically prohibits this move because it would damage confidence in public finance, but it would be possible if approved by the Diet. This could be an emergency measure along the lines of a necessity in a crisis. In a case like this, however, the government would have to prepare for the market's reaction. Because confidence in the Japanese government and the yen would immediately plummet, the trend to sell yen and buy other currencies would accelerate, triggering a severe depreciation in the national currency and increased inflation. A sharp depreciation in the yen would mean acute volatility in other currencies such as the dollar and euro, greatly affecting the world economy. In particular, if long-term interest rates soared due to rapid inflation, there is a strong likelihood that government bonds would slump, leading domestic financial institutions with huge government bond portfolios to suffer unrealized losses, which would push the domestic economy and financial system further into chaos. This, in turn, would throw the global financial system into turmoil. The impact would be far greater than the subprime loan exposure that triggered the 2008 global financial crisis.

The banks are the ones that would be placed in a predicament here. There is a risk that depositors seeking other currencies and better returns would withdraw their deposits across the board. If possible, banks would raise interest rates to prevent this but if they did that the banks themselves, who manage a large share of their funds in government bonds, would end up with "reverse margins." Management would be quickly compromised. Other companies, of course, would not go unscathed. A rise in interest rates means even heavier corporate debt. Loss-cutting measures such as restructuring employees and curbing investment in order to give priority to servicing debt are likely. But if they are unable to do so, companies would go bankrupt. If all of these scenarios occur, the impact on households would be enormous. Family breadwinners might lose their jobs, and if not, the possibility of pay cuts would be enormous. On the other hand, if prices rose steadily through inflation, life would certainly become more and more difficult. That would be acceptable if the situation settled down in the short term, but the chance of longer-term protraction is great. If things truly came to this point then the government would probably implement measures of last resort. Since deteriorating public finances originally triggered inflation, it would no doubt implement drastic fiscal reform to reassure both domestic and foreign interests. However, pensions, medical insurance, as well as defense, would need to be cut and coupled with a significant tax increase. Should all this be implemented as the economy cooled down, it would merely add insult to injury for household budgets. There might even be an increase in the number of families abandoning Japan and escaping their predicament by moving abroad.

The above scenario is a very one-sided one and it is difficult to predict with any accuracy what would really take place. But one conclusion is very clear: if fiscal and social security reform are not tackled, many Japanese, if not portions of the world economy, will be caught up in a great and ensuing chaos.

Notes

1. Cabinet Office 2015.
2. Cabinet Office 2014; Nikkei 2014.
3. For example, in its *Global Japan: 2050 Simulation and Comprehensive Strategy* (2012), the Twenty-First Century Public Policy Institute predicts "there will be minus growth in all scenarios after the 2030s" (Keidanren 2012).
4. Act submitted June 20, 2012, passed August 10, 2012 (Cabinet Office 2010).
5. Prime Minister of Japan 2014.
6. In the Ministry of Finance's *Provisional calculations for the cash flow situation in the Government Bond Consolidation Fund* (MOF 2015, February), interest payments are expected to swell to about 25 trillion yen a year in nine years' time up from about 10 trillion yen currently (Kawamura 2011).
7. Fiscal drag is the (endogenous) effect of changes in economic activity and incomes on tax revenues due to the progressivity of the tax system.
8. The government had run up huge debts in the wartime economy against the private sector through contracts and orders to businesses. This relationship became the issue of wartime debt compensation when Japan surrendered on August 15, 1945 in the Pacific War (Takahashi 2012).
9. Bank of Japan 2016.
10. IPSS 2013.
11. Medical services that can be received in conjunction with treatment covered by insurance by insurance subscribers bearing an additional cost.
12. General Government Net Saving = Current revenue such as tax minus current expenditure.

References

Bank of Japan. (2016, December 19). Basic Figures: Flow of Funds for the Third Quarter of 2016 (Preliminary report, Research and Statistics Department). https://www.boj.or.jp/en/statistics/sj/sjexp.pdf, accessed on February 13, 2017.

Braun, R. Anton & Joines, Douglas. (2014). The Implications of a Graying Japan on Government Policy (mimeo, Federal Reserve Bank of Atlanta). http://www.frbsf.org/economic-research/files/5Braun-Joines.pdf, accesses on February 13, 2017.

Cabinet Office, Government of Japan. (2010, December 14). Promotion of Social Security Reform (summary of Cabinet Decision). http://www.cas.go.jp/jp/seisaku/bangoseido/pdf/english/dec142010.pdf, accessed on September 25, 2016.

Cabinet Office, Government of Japan. (2014, November). Choice for the Future: Creating a Growth and Development Model Made in Japan Transcending Rapid Depopulation and Super Aged Society (white paper, Council on Economic and Fiscal Policy, Expert Panel, Committee for Japan's Future). http://www5.cao.go.jp/keizai-shimon/kaigi/special/future/houkoku/01_e.pdf, accessed on September 25, 2016.

Cabinet Office, Government of Japan. (2015). Annual Report on the Aging Society: 2015 (white paper). http://www8.cao.go.jp/kourei/english/annualreport/2015/2015pdf_e.html, accessed on September 25, 2016.

Fukao, Kyoji. (2010). Two Lost Decades and Japanese Economy (Ushinawareta 20nen to nippon no keizai). *Research Institute of Economy, Trade and Industry, Research Digest no. 3*, 23–28. http://www.rieti.go.jp/en/publications/rd/003.html, accessed on September 25, 2016.

Hansen, Gary D. & Imrohoroglu, S. (2016). Fiscal Reform and Government Debt in Japan: A Neoclassical Perspective. *Review of Economic Dynamics* 21. doi:10.1016/j.red.2015.04.001, accessed on February 14, 2017.

Ikegami, Naoki. (2006). Should Providers be allowed to Extra Bill for Uncovered Services? Debate, Resolution, and Sequel in Japan. *Journal of Health Politics, Policy and Laws*, 31(6), 1129–1149. http://jhppl.dukejournals.org/content/31/6/1129.abstract, accessed on September 25, 2016.

IPSS. (2013, October). Financial Statistics of Social Security in Japan 2013 (Shakai hosho hiyo tokei). http://www.ipss.go.jp/ss-cost/j/fsss-h25/H25.pdf, accessed on September 25, 2016 (in Japanese).

Kawaguchi, Hiroyuki. (2012). Unique mixtures of public and private funding in Japan: The ban on billing for mixed medical care service. *MOF Public Policy Review*, 8(2), 145–170. https://www.mof.go.jp/english/pri/publication/pp_review/ppr016/ppr016b.pdf, accessed on February 2, 2017.

Kawamura, Sayuri. (2011). The outlook for future government bond issuance and fiscal management in Japan (Wagakunino kokusai hakko to zaiseiun-ei no sakiyuki), *Business & Economic Review*, December 25, 2011, https://www.jri.co.jp/page.jsp?id=20474 accessed on February 14, 2017 (in Japanese).

Keidanren. (2012, April 16). Global Japan: 2050 Simulation and Comprehensive Strategy (21st Century Public Policy Institute). http://www.21ppi.org/english/pdf/120827.pdf, accessed on October 2, 2016.

MOF. (2015, February). Provisional calculations for the cash flow situation in the Government Bond Consolidation Fund (data table). http://www.mof.go.jp/budget/topics/outlook/sy2702b.htm, accessed on October 6, 2016 (in Japanese).

Nikkei. (2014, November 14). Negative 0.1 growth if no actions taken—report from the Committee for Japan's Future (Jinkogen hochi nara mainasu 0.1% seichouni seifui houkoku). *Nikkei Shimbun*. http://www.nikkei.com/article/DGXLASFS14H2G_U4A111C1PP8000/, accessed on October 5, 2016 (in Japanese).

Oguro, Kazumasa. (2014, April 7). Inaccurate economic target and unrealistic growth prediction (Seichono minamoto wa machigatta keizai mokuhyo). *Nikkei Business Online*. http://business.nikkeibp.co.jp/article/topics/20140403/262300/?P=1, accessed on September 25, 2016 (in Japanese).

Oguro, Kazumasa. (2015, December 14). Cost analysis: relationship between medical expenses and actual diseases (Iryohi no jiko futan wo shippeibetsuni). (Canon Institute for Global Studies). http://www.canon-igs.org/column/macroeconomics/20151214_3403.html, accessed on September 25, 2016 (in Japanese).

Oguro, Kazumasa. & Satoshi Shimasawa. (2011). *An Introduction to Macro-Economic Modeling Using MathLab (MathLab ni yoru makuro keizai moderu nyumon)*. Tokyo: Nippon Hyoron sha Co., Ltd.

Prime Minister of Japan and His Cabinet. (2014, November 18). Opening statement by Prime Minister Shinzo Abe (Abe naikaku sori daijin kishakaiken). http://japan.kantei.go.jp/96_abe/statement/201411/1118kaiken.html, accessed on September 25, 2016.

Takahashi, Masahiko. (2012, May). The History and the Basic Issues of the Deposit Insurance System (Yokinhokenseido no rekishi to kihontekikadai), *Deposit Insurance Research,* May issue, https://www.dic.go.jp/katsudo/chosa/yohokenkyu/201205-14/14-1.html, accessed on September 25, 2016 (in Japanese).

CHAPTER 5

Demographic Policy

Inagawa Hidekazu

One can describe Japan's demographic policy as doing too little and coming too late.[1] As a result, the population is steadily falling (Chino 2009). Data from the Ministry of Internal Affairs and Communications (MIC 2015) and the Cabinet Office (2013) show that the speed is nothing like that of Europe. At the end of the eighteenth century, the British economist Thomas Malthus noted in *An Essay on the Principle of Population* that, if left unchecked population would increase exponentially.[2] Today, however, the opposite phenomenon is taking place. The government's reluctance to intervene in the population question and its inability to predict a falling birth rate and drop in the population, along with its weak sense of crisis concerning the graying of Japan, are major factors in its failure to implement an effective demographic policy. The failure to set a national agenda on the issue of immigration, a factor that could potentially hold the key to resolving the population problem, also has had an impact.

Inagawa. H (✉)
Faculty of Sociology, Toyo University, Tokyo, Japan

Y. Funabashi (ed.), *Japan's Population Implosion*,
https://doi.org/10.1007/978-981-10-4983-5_5

5.1 Japanese Demographic Policy: Between "Too Much" and "Too Little"

Taking a broad look at the state's attitude toward the population question, we can see that before World War II until the immediate aftermath, Japanese demographic policy wavered between a theory of "under-population" and one of "overpopulation." After the 1918 rice riots and the worsening economic situation in the early 1930s known as the Showa Depression, the overpopulation theory dominated political discourse.[3] This is apparent from the Investigation Committee on the Population Food Problem established in 1927, which encouraged emigration. However, with the outbreak of the Sino-Japanese War in 1937 a policy premised on the under-population theory appeared on the scene due to the growing need to secure military troops.[4] The 1940 National Eugenics Law, which aimed to increase the number of "physically and mentally sound people" as defined at that time, and the 1941 Guidelines for Establishing a Population Policy, symbolized by the slogan of "Beget and Multiply," are examples of this. These policies were obviously designed to encourage population growth. Nevertheless, after the war the theory of overpopulation once again took hold. A key factor here was the social fear about food security, for example, as growing numbers of demobilized soldiers and millions of repatriated Japanese returned from abroad to rural areas. The Eugenics Protection Law (enacted in 1948 and which permitted abortion) stated that maternal protection was its purpose, but the overpopulation theory also acted as an incentive for the promotion of abortion as one solution to increasing birth rates. In 1954, the Council on Population Problems issued "Resolution on the Quantitative Adjustment of the Population," thereby launching a general policy of population control.

The theory of under-population quickly grew outdated among these trends, and most politicians and bureaucrats focused on Japan as an over-populated country. Entering the high-growth period of the 1960s, a shortage of labor was also deemed a problem but this was dealt with by encouraging large numbers of workers to migrate from Japan's rural areas.[5] Before long, Japan experienced its own postwar baby boom, a "demographic bonus," and the ratio of the working generation to children and the elderly grew substantial. This boom obscured the potential demographic problem. There was no awareness of the need to prepare for a future population decline, and the under-population theory ceased to appear prominently in national policy debates.

Demographic focus subsequently switched to how to shift from explosive growth to the maintenance of steady population levels. In its 1974 report, "Japanese Demographic Trends: Aiming for Static Population," the Council on Population Problems expressed the view that an absence of either population increase or decrease was desirable.[6] The goal was not to increase the population by switching to a "low-birth, low-death type" after transitioning through a "high-birth, high-death type" and a "high-birth, low-death type" (primary demographic transition). In reality, by 1975—which could be referred to as the end of the high-growth period—the total fertility rate (TFR) had already fallen below the replacement level. The rate continued to decline thereafter, only making a brief recovery in the early 1980s. This brief upward spike played a part in the decline in the TFR not being recognized as a long-term trend as well as in the failure to consider population decline as a problem, as we noted in Chapter 1. By the late 1980s, however, the TFR once again plummeted. Experts finally recognized that the downward trend was not a temporary one. In 1988 the Council on Population Problems mentioned the low birth rate issue for the first time. It pointed out that changes in family structure had led to the declining birth rate, such as the shrinking number of extended families and rise in nuclear families, as well as an increase in working couples.

Japan was rocked by the "1.57 Shock" in 1990, when the TFR had dropped to 1.57.[7] This message at once ignited greater public interest in population decline, and the government also sharpened its crisis awareness toward the problems that a declining birth rate would create for the nation. Prime Minister Kaifu Toshiki's policy speech in March 1990 stated that a decrease in births is "posing a variety of problems for the future of our country," and he announced strengthening measures for childbearing and rearing.[8] The 1992 White Paper on the National Lifestyle, entitled "Impact of and Response to the Advent of a Low Birth-Rate Society," dealt with the declining birth rate problem.[9] In 1994, the government drew up the "Angel Plan" to combat declining birth rates and the "Five-Year Project on Emergency Childcare Measures," improving the childcare environment. "Basic Policy on Promoting Countermeasure to Declining Birth Rates" and a "New Angel Plan" was established in 1999, which recalculated goals and added new measures. In 1997, when the nursery school reform's top issue was its admission system, a new system was introduced to correct this issue that municipalities took over the role for applications. Mori Yoshiro became the first prime minister to declare in a policy speech that the Japanese government would squarely address its declining birth rate in January 2001.[10] In United Nations surveys on population issues, Japan answered in 1996 that, while the TFR was too low, "the government would not intervene." However, after 2003, Japanese officials announced that the country would "maintain or raise" the level of the population.

5.2 Raising Children: A "Social Responsibility" or a "Personal Issue"?

Japan's policy responses to the country's declining birth rate have amplified over the years, but it remains a serious problem. This is because given the speed of the fall in the birth rate, measures have been too late and in terms of content, they have not had a great enough impact to push a bold and radical set of reforms.

Why did this happen? First, countermeasures for a declining birth rate are a sensitive issue. Given that dating, marriage, pregnancy, childbirth, and child-rearing are closely associated with individual values and choices, it is difficult to create a social consensus. Remorse over the prewar policy of "Beget and Multiply" also played a role and people have a strong psychological resistance toward political and government intervention in the population question. Various arguments also exist concerning parenting. A variety of services are required for children in their early years. The "three-year-old myth"—which theorized a deleterious influence on child development if mothers did not raise children at home up until 3 years of age—remains a potent social idea. Although children are referred to as "society's treasures," parenting tends to be perceived as an individual issue. A question that plagues municipal authorities regarding custody in every case of child abuse is whether the possibility of being able to save children increases if the idea that children "belong to society" is emphasized; but opposition to such an approach is not insignificant (Ueki 2011).[11] Therefore, the declining birth rate involves not only individual values but also social values and norms. For example, Japanese public opinion is divided about whether raising children is a "parental responsibility" or a "personal choice." The debate also touches on questions where the discussion can easily lead to conflicts such as those over abortion and reproductive health/rights.

If the government attempts to implement bold measures in a situation where a convergence of values is difficult, they will be met with opposition from either one side or the other. Therefore, under a minimal consensus, the government has cautiously gone ahead making policy from the assumption that the primary responsibility of parenting lies with parents but with the support of society, so as not to rock the boat or impinge on values.

5.3 Demographic Policy Behind the Curve

The state's overly optimistic take on demographic change is another background factor in its failure to derive a set of timely and effective policies. Unlike economic forecasts, there is a certain degree of accuracy to demographic projections. From 1955 to 1964, the Ministry of Health and Welfare, now the Ministry of Health, Labour, and Welfare (MHLW), worked on enacting elder welfare legislation, creating the pension scheme and separating elder welfare from the Public Assistance Act.[12] The government correctly foresaw that Japan would one day face the question of aging. (Such was the weakness of public awareness of the crisis that the pension scheme met with much opposition in the National Diet and elsewhere.) However, the state was unable or unwilling to predict the drop in population associated with a low birth rate. Demographers assumed the population would remain constant because they were under the influence of "Demographic Transaction Theory" (the theory that a population transitions through three stages, from "high-birth, high-death" to "high-birth, low-death" to "low-birth, low-death"). Accordingly, it was not easy to introduce the current mainstay of countermeasures for declining birth rates—introducing child allowances or reducing the economic burden of child-rearing—but Japan finally institutionalized them in 1971 after ratifying the international convention on minimum standards of social security by the International Labor Organization.[13] The child allowance was supposed to start small and grow larger, but has been relatively ignored until recently with calls for its abolition emerging every so often.

Currently, the Statistics Bureau announces the Population Census once every 5 years. Since this timing is also linked to the fiscal recalculation of pensions (the current fiscal verification), a drop in the birth rate usually only comes into question depending on how it will affect the sustainability of the pension scheme. This too may well have blunted social and political interest toward predictions of depopulation. It was only from the late 1980s that experts recognized the drop in the birth rate was not transient. Demographic policy to date has been biased toward social security and employment policy (childcare support and work-life balance measures). A drop in the birth rate is an issue that concerns the whole gamut of national government, including education, housing, transportation, and the tax regime. Yamaguchi Kazuo, a professor at the University of Chicago, cites several factors preventing an increase in the birth rate. Professor Yamaguchi notes, in the case of the first child, "a lack of work and family balance"; for the second child, "the low level of paternal participation in childcare"; and for the third child, "the magnitude of the economic burden of child support and education costs" (Yamaguchi 2013). Even government surveys have shown that it is easier for a second child to be born in families where the husband does housework and child-rearing on his days off, and that women raising children believe "economic support" to be the most important countermeasure for the declining birth rate.[14]

However, when it comes to actual measures, the government's focus remains set on social security and employment policy. Even in the 2003 Basic Act for Measures to Cope with Society with Declining Birthrate, which declared it would address the low birth rate in a comprehensive manner citing the "unprecedented nature of the situation," the emphasis was largely on employment and social security. For example, its eight basic policy items included development of the employment environment, enhancement of childcare services, and creating childcare support systems in the community.[15] This policy bias facilitated lifestyles where even if childcare leave were available it was impossible to use, and social values that inhibited career building. A further background factor was the unwillingness of politicians and the relevant ministries to introduce measures for the population problem due to fears of a public backlash. The Basic Act for Measures to Cope with Society with Declining Birthrate comes under the jurisdiction of the Cabinet Office. As a consequence, the Cabinet Office has developed various approaches in cooperation with the relevant ministries and agencies such as the MHLW and Ministry of Finance (MOF). Nevertheless, the population problem is an area that is difficult to coordinate for government bodies such as the Cabinet Office, whose role is mainly a regulator. As a result, declining birth rate countermeasures are laid out irresponsibly without securing the necessary budgetary resources.

5.4 The Cold Realities of Childcare Support

Even the content of childcare support measures, where the greatest policy emphasis has been placed, is inadequate and improvements have not progressed smoothly. A postwar child benefit was first broached in the 1947 Outline of the Social Security System of the Social Insurance System Research Council and the Council on Population Problems proposed its establishment in 1962. But in light of the atmosphere of the time, childcare was associated with population growth threats and prewar birth incentives, so child allowances were quietly forgotten. After fits and starts, the measure was eventually implemented in 1972. In direct contrast to initial fears of population growth, the scheme was launched in the era's climate of a falling birth rate. Sadly, the child allowance did not progress as planned. Voices were raised from companies—who were to bear some of the financial burden—that child-rearing expenses were already covered by wages, and that it was therefore a double burden. Since public funds were to be injected, and before the demographic issue became viewed as the problem it is today, time and again the MOF and the wider community called for a review of the allowance as a waste of public resources.

Following this rather desultory debate on child benefits, the financial resources injected into the child allowance scheme today have increased significantly. Although income testing is severe, the amount has been raised to a level not inferior in comparison with other developed countries such as France. When the scheme was first introduced in 1972, the level of the child allowance (3000 yen/month) was almost equivalent to the non-contributory old-age welfare pension (3300 yen/month). Today, the child allowance (10,000 or 15,000 yen/month) is far below the level of old-age welfare pension (33,308 yen/month). This gap is due to adjusting eligibility and the allowance amount when there is no significant increase in financial resources until the revision of the child allowance, such as keeping the total amount the same but raising the allowance level and reducing the eligibility age or limiting eligible children. The state has expanded payment from the third child to the second and further to the first child, but it also curtailed the eligible ages or allowance period as well as tightening up income testing.

Nurseries currently play a greater role supporting working couples than they do for the impoverished, although their origins are found in measures for poor factory workers and farming households during the busy seasons. It is for this reason that facilities for childcare cannot keep up with greater demand and long waiting lists exist in urban areas. As a result, unauthorized nurseries have emerged where the services are not necessarily high enough to meet actual childcare. One policy taken up in the 2000s as a necessary part of regulatory and structural reform based on this situation was "pre-school/kindergarten integration," where *crèches* were to be incorporated into kindergartens. In 2006, the Act on Certified Children's Preschools was enacted to allow new facilities combining both these functions. However, the integration of preschools and kindergartens is not going smoothly due to a variety of differences.

Under the 1947 Child Welfare Act, preschools were defined as a child welfare facility responsible for childcare (education and nursing) catering for infants from zero years of age. But based on the School Education Act of the same year, a kindergarten is a school (so-called Article 1 school) responsible for early childhood education catering for toddlers more than 3 years old. Although there are overlaps in function and age, their systems and regulations are different. Staff status and regulations also differ between public preschools and kindergartens. Nursery personnel are part of the welfare service while kindergarten personnel belong to the educational service as schoolteachers, which means that the latter are subject to restrictions under labor law that are not applicable to public servants in regular service. Different systems of social insurance and pensions in private preschools and kindergartens also further exacerbate divisions. In terms of unions, public kindergartens belong to the Japan Teachers' Union and public nurseries to the All Japan Prefectural and Municipal Workers' Union. Politicians can also be divided into two competing groups based on their attitudes toward the child allowance. Because Diet members each support one group or the other, it is difficult to implement a reform of the system where the interests of the two groups are in conflict. Labor-oriented representatives have close ties with preschools, and Diet members with interests in education share close ties with kindergartens. Preschools fall under the jurisdiction of the MHLW, while kindergartens come under the aegis of the Ministry of Education, Culture, Sports, Science and Technology (MEXT). These legal and ministerial differences have created diverse conflicts of interest in preschool/kindergarten integration, impeding reform in childcare administration. The interests of the end users—children and parents—have not been given top priority.

5.5 Still Unresolved Policy Response on the "Foreign Workers" Issue

Up to this point, we have looked at demographic policy from the perspective of the declining birth rate. But the population problem is also a labor force problem. The Japanese government has set no broad direction concerning what to do with non-Japanese workers, whose intake is only being somewhat expanded. Partly due to the far-ranging impact of foreign workers, such as the perceived impact on wages, the education system, and security, the situation remains one of endless discussion that never arrives at any clear policy decision that convinces all stakeholders. As such, progress has only been made in the employment of overseas residents of Japanese ancestry (mainly from Brazil and Peru) and short-term technical trainees. As things stand, the government has not been able to respond to the various social issues that will most likely arise with the immigration of foreign nationals. The number of foreigners resident in Japan at the end of 2013 was approximately 2.07 million. Although this is a drop from the 2008 year-end peak of 2.14 million, it is a 1.6% increase on the 2.03 million at the end of 2012. The share by nationality is China (31.4%), South Korea/North Korea (25.2%), the Philippines (10.1%), and Brazil (8.8%).[16]

Japan has a low population ratio of foreigners compared with other developed countries. According to the OECD *International Immigration Outlook 2013*, the 2011 proportion of foreign-born people is 12.6% on average in the member states where data were taken, with 10 countries having more than 15%. Japan does not even reach 2%.[17] According to reports mandated by the Employment Measures Act, foreign workers number approximately 790,000 as of the end of October 2014.[18] However, this figure includes some 340,000 people with special status (people of Japanese descent, permanent residents, and those with Japanese spouses), approximately 150,000 are in skills-training programs, and a further 150,000 are engaged in activities other than those permitted under their previously granted status of residence. Only 150,000 people were admitted for employment purposes.

The Japanese government has looked into accepting foreign workers and dividing them into "highly skilled foreign professionals" and "unskilled workers." Regarding the highly skilled, the Basic Employment Measures Plan and Economic Plan, formulated between 1988 and 1999, defined the active promotion of an intake of professional and technical foreign workers to further Japan's internationalization and revitalization of the economy.[19] The government introduced a points system from 2012 where acceptance is determined by a value system assigned for each occupation and educational background. This is based on the idea that high-skilled workers are essential for Japan's development. The Immigration Control and Refugee Recognition Act (Immigration Control Act) was revised in 2014 and shortened the period of stay required for permanent residency acquisition from 5 to 3 years. This plan also envisioned the long-term residence of highly skilled foreign nationals. The government has also promoted domestic employment support for exchange students in professional and technical fields. However, there is also international competition for advanced human resources.

In the previous two plans the government stance was to proceed cautiously because of the tepid national acceptance of unskilled workers; in fact, the Immigration Control Act was not set up to accept unskilled workers. The status of residence for foreign nationals entering Japan is limited to professional and technical fields and there is no category that corresponds to "unskilled labor." In reality, however, although there has been a steep need for foreign workers in professional and technical fields, this is even more so the case in the unskilled fields of manufacturing and agriculture. Japan currently faces a situation where foreign workers in the form of "technical trainees" and "foreign residents of Japanese descent" are employed to meet the needs of sectors that cannot be covered by domestic workers alone. "Technical trainees" refers to foreigners who enter Japan through a skills training system created for the purpose of achieving a transfer of technology to developing countries. Through the intervention of small business organizations in 1991, trainees were permitted to assist small and medium-sized enterprises that were suffering labor shortages. In the 2010 system reforms, the trainees' residence status, previously categorized as a "special activity," was changed to "skills training" after completion of the training period, and labor law and regulations are now being applied to them. Foreign workers of Japanese descent are people with Japanese lineage and their families, such as from Brazil and Peru. No limit was placed on their employment activities and their numbers have increased rapidly since 1988 since they are mainly given long-term residence status, and they are engaged primarily in manufacturing in regions suffering a shortage of workers (Goto 2012).[20]

In February 2014, the government's Committee for Japan's Future estimated that a significant decline in population over the next 100 years could be avoided by allowing an annual intake of 200,000 immigrants.[21]

At nearly the same time, the Liberal Democratic Party's Headquarters for Japan's Economic Revitalization issued a recommendation for accepting foreign workers, including extending the skills training system from 3 to 5 years. A Cabinet meeting also looked into expanding the skills training intake for the purpose of meeting the 2020 Tokyo Olympics construction demand.[22] Additionally, various deregulations promoted in the National Strategic Special Zones, have seen a relaxation of visa rules as incorporated into the Global Start-Up National Strategic Special Zone in Fukuoka.[23] Reflecting this trend, the May 2014 summary of the Committee for Japan's Future's interim report, *Choice for the Future*, stated that "in order to increase growth potential in a declining population, exchange with foreign workers will be used for growth and development through their strategic acceptance—including advanced high skilled workers—while promoting a national debate on this."[24] The June Basic Policies for the Economic and Fiscal Management and Reform 2014 adopted by the Cabinet described the use of foreign workers as "internal globalization." Even these, however, note that the use of foreign workers is not an immigration policy, which means that Japan at present is still not moving ahead in its fundamental thinking about immigration as either necessary for its long-term survival or as a countermeasure to depopulation.

5.6 A Seclusion or Open Door Policy?

Japan's position on accepting foreign workers is stymied because the government has struggled in the face of conflicting stakeholder stances. The main stakeholders in the foreign workers issue are the business community, labor, and local governments. The business community was previously reluctant to accept foreign workers. However, it shifted to active support in 2000, around the time of the 2001 publication of the *Labor Studies Committee Report* by the Japan Federation of Employers' Associations (*Nikkeiren*). In 2004, the Japan Business Federation (*Keidanren*), recommended comprehensive intake measures to increase value added creativity rather than a disorganized intake to make up for Japan's population decline.[25] The report suggested that advanced human resources and workers in service areas with labor shortages, such as welfare, as well as workers in primary industry sectors, such as agriculture, forestry, and fisheries, were necessary. By contrast, organized labor has taken a more cautious approach. In 1988, the Japan Private Sector Trade Union Confederation reconfirmed its policy in *Japan Private Sector Trade*

Union Confederation's Standards on Prerequisite and Criterion for Accepting Foreign Workers that limited intake of non-Japanese workers to occupations that require specialized knowledge, technology, and skills. (Japan Private Sector Trade Union Confederation 1988). The subsequently formed Japanese Trade Union Confederation maintained this position, and currently asks that residence status and work eligibility for aliens not be deregulated to allow unskilled labor, and that the use of Japanese female and elderly workers should take precedence over any discussion of foreign worker intake.

Local municipalities that are already facing issues with foreign residents of Japanese descent are in a position of constantly having to respond to their long-term needs.[26] In 2001, the heads of local governments who deal with foreign residents of Japanese descent established a "Meeting for Cities with Foreign Communities."[27] The passive attitude of the relevant ministries and agencies to the foreign workers issue also slackens the approach of the government as a whole. Within the government, it is the Cabinet Secretariat that is in charge of the foreign workers issue. However, different ministries respond in individual areas of immigration during and after the actual immigration. For example, the MOJ, Ministry of Foreign Affairs (MOFA), MEXT, and MHLW mainly handle border control, the first port of call when foreign workers immigrate. After entry, almost every ministry or agency is involved, including the police (for crime), MEXT (education), MHLW (social security), Ministry of Land, Infrastructure Transport & Tourism (MLIT) (Housing), and MIC (local governments). The Ministry of Economy, Trade and Industry (METI), which is being lobbied by the business community to accept an expansion of foreign workers, has displayed an active desire to accept some workers. However, as almost all the other ministries do not want to deal with this complicated business, the issue has lapsed and stagnated.

Debate over immigration in terms of maintaining Japan's "seclusion" or "opening the door" is currently a hot topic in Japan. Historically, it tends to be forgotten that Japan has both accepted immigrants and dispatched emigrants. In previous times, there was interaction with the Asian mainland by bringing people over, and when the population was growing in the modernization of the Meiji Era (1868–1912), emigration was promoted as a national policy. Food shortages related to population growth have acted as a catalyst for migration, including the Irish immigration to the USA. We cannot deny that the movement of people has created many conflicts and been the source of drama. The question of immigrants and foreign workers relates to the very core of the nation and should not be conceived of in economic terms alone. A socio-political viewpoint that is more comprehensive and includes religion, culture, and ideology is also required, as is a composite approach that focuses not only on the home country but foreign countries as well. In considering Japan's uninspired immigration policy, there is much to be learned from history. What is especially important is to realize that the impact of cross-border movements once made remain throughout history. Even if policy is interrupted, the effects of that policy remain. This can be seen in Japan's prewar colonial settlement of Manchuria and postwar emigration to the Dominican Republic. In this regard, the movement of people is very different from, let us say, the international trade of short-term consumer goods.

To date, Japan's immigration policy has oscillated between believing in both overpopulation and under-population theories. Just as is the case with demographic policy, immigration policy is by nature an area that requires a sustained political policy approach and not mere piecemeal accommodation.

Notes

1. "The Japanese population finally went into decline last year. The time has come to focus on the problem, though it feels a little late to be doing so." (Daiki 2006).
2. In the first edition of his *Essay on the Principle of Population* published in 1798, Thomas Malthus of the United Kingdom expounded his famous hypothesis: "Population, when unchecked, increases in a geometrical ratio, subsistence, increases only in an arithmetical ratio' and '[t]superior power of population cannot be checked without producing misery or vice." (Malthus 1798).
3. Kito 2006; Asahi Shimbun 2008. The mid-Taisho to early Showa years brought a sense that Japan was overpopulated. The rice riots of 1918 are emblematic of this era. The self-sufficiency rate for rice fell from 94% in 1915 to 83% in 1939 due to the uneven development of agriculture and manufacturing. The promotion of rice imports only accelerated the deterioration of the agricultural communities. The government responded to this by systematic emigration.

   "The year is coming to a close as the financial crisis than began in the United States spreads through the rest of the world. It was the same at the year's end in 1929. A stock market crash in the New York Stock Exchange had dragged the United States into the Great Depression, and Japan had also plunged into its own Showa Depression." The Showa Depression and the Great Depression - Economic growth rate (nominal): Japan-18.3%, Unites States-49.7%."

4. Hiroshima 2006.
5. Mainichi Shimbun 2016. Yamaguchi Satoshi's 2013 book *What Was Getting Jobs En Masse All about? The Space-Time Continuum of the Golden Eggs* (*Shudan shushoku towa nande atta ka—kin no tamago no jikukan*) explains the definition of "getting jobs *en masse*" as "mainly the mass relocation of young labor, mainly new graduates from middle school, and related phenomena caused by a range of systems of public organizations during the post-World War II high-growth era." The period of Japan's rapid economic growth during 1955–1973 is generally known as the high-growth era.
6. The Council on Population Problems 1974.
7. Nikkei 2015. The catchphrase "1.57 Shock" swept like a whirlwind through Japan in 1990. It was revealed that the "year of the Fire Horse" superstition had caused the previous year's total fertility rate to dip below the previous post-World War II low in 1966.
8. "The decline of the birth rate raises many questions about the future of our country. We must entrust Japan's future to actively supporting the willingness of young people to have children and make efforts to that end. Children are the treasures of our world. To protect this treasure and to enable it to grow healthy and robust is more important than any other task. I shall aggressively promote the creation of an effective environment with these thoughts engraved on my heart (National Diet Library 1990)."
9. Economic Planning Agency 1992.
10. "The rapid deterioration in birthrates in recent years has engendered concern that there will be a widespread influence on the economy and society of our nation. It is important that the government work together to take comprehensive measures in order to ensure that in the 21st century Japan remains a society in which our people can have dreams and hope for their families and for raising children in a bright future. That is why, based on the Basic Policy on Advancing Measures in Response to the Falling Birthrates, I will submit to this Session of the Diet bills to revise the Law Concerning the Welfare of Workers Who Take Care of Children or Other Family Members, Including Child Care and Family Care Leave in order to advance the creation of an employment environment to facilitate a society in which people can give birth to and raise their children while continuing to work. At the same time, I will try to enhance childcare services through steps such as expanding the framework for accepting younger children at daycare facilities. In this way, I will advance comprehensive measures to respond to falling birthrates spanning broad sectors including social welfare, employment, education and housing." (MOFA 2001)

11. "People have pointed to the need to review the parental authority provisions in the Civil Code because of such matters as the existence of persons who have parental authority who try to justify child abuse or make unjust demands regarding institutionalized children under the pretext of parental authority and the reluctance towards petitioning or ruling for loss of parental authority due to the fact that the entire parental authority is taken away without a time limit under the parental authority loss system" (Ueki 2011).
12. Ministry of Justice (MOJ) 1950.
13. ILO 1952.
14. MHLW 2005; Cabinet Office 2008.
15. Other areas not mentioned in the text include enhancement of the maternal and child health medical system, promotion of more relaxed education, improvement of the living environment, reduction of economic burdens, and education and enlightenment.
16. MOJ 2015.
17. OECD 2013.
18. The law was enacted in 1966. Its purpose is to create employment security for workers and improving workers' economic and social status, as well as to contribute to economic and social development and the achievement of full employment (Employment Measures Act 1966).
19. 6th Plan of 1988; 7th Plan of 1992; 8th Plan of 1995; 9th Plan of 1999. 5 Year Economic Plan: Japan Living with the World, 1988; 5 Year Plan for a Life Superpower, 1992; Socio-economic Plan for Structural Reform, 1995; Policy Plan for an Ideal Economy and Economic Rebirth, 1999.
20. Number of Brazilian immigrants—under 10,000 in 1988; over 230,000 in 2010. Number of Peruvian immigrants—under 10,000 in 1988; over 52,000 in 2010 (Goto 2012).
21. Cabinet Office 2014a.
22. Agency for Cultural Affairs 2015.
23. A zone designated by the national government based on the perspectives of boosting the international competitiveness of industry and promoting the creation of centers of international economic activities by giving priority to advancing structural reform of the economic system. Tokyo Metropolis, Kanagawa Prefecture, and Narita City in Chiba Prefecture have been designated as the Tokyo Area zone by the national government (Tokyo Metropolitan Government 2013).
24. "So as to lift the growth potential under the decreasing population, while it is important to heed the national debate, a commitment should be made to accept strategic immigration of a highly skilled foreign labor force and apply the mutual exchanges with them to complement economic growth and development" (Cabinet Office 2014b).
25. See Nikkeiren (2001), and the recommendations on the question of accepting foreign workers in Keidanren (2004).

26. Although the Japanese government created an action plan in 2011 on measures for long term residents of Japanese ancestry centered on education and everyday life support, there is no end to foreigners of Japanese ancestry leaving Japan because of discriminations and other reasons. The Brazilian population of the city declined by more than 10% in 2013, even though the economy had improved (Nikkei Veritas 2014).
27. Efforts have been made to improve multilingual adaptation of portal sites for long term resident foreigners. The Congress of Cities with Concentrated Population of Foreigners consists of the administrations of cities and regional international exchange associations with large populations of Latin Americans of Japanese ancestry known as "newcomers" and other foreigners. It was established to exchange information regarding measures and activities concerning resident foreigners in order to actively work to resolve the various challenges that are emerging locally.

References

Agency for Cultural Affairs. (2015). Regarding the acceptance of foreign labor (Seifu zentai ni okeru gaikokujinzai no ukeire katsuyo ni kansuru ugoki nado). http://www.bunka.go.jp/seisaku/bunkashingikai/kokugo/nihongo/nihongo_66/pdf/sanko_4.pdf, accessed on September 25, 2016 (in Japanese).

Asahi Shimbun. (2008, December 25). Learn from the Showa Depression, 1929, from USA (1929-nen Beikoku-hatsu showa kyoko ni manabe). *Asahi Shimbun (morning edition)*.

Cabinet Office. (2008). Awareness survey of child-raising women on declining birth rate society measures, Chapter 5, 43. http://www8.cao.go.jp/shoushi/shoushika/research/cyousa20/ishiki/pdf/02chap_5.pdf, accessed on September 25, 2016 (in Japanese).

Cabinet Office. (2013). Declining Birth Rate White Paper: 2013 (Heisei 25nen-ban shoshika shakai taisaku hakusho). http://www8.cao.go.jp/shoushi/shoushika/whitepaper/measures/w-2013/25webhonpen/html/b1_s1-1.html, accessed on September 25, 2016. (in Japanese).

Cabinet Office. (2014a). Abstract from the Third Choice for the Future Meeting (Giji yoshi). http://www5.cao.go.jp/keizai-shimon/kaigi/special/future/0224/gijiyoushi.pdf, accessed on September 25, 2016.(in Japanese).

Cabinet Office. (2014b). Council on Economic and Fiscal Policy, Expert Panel, Committee for Japan's Future. Choice for the Future—Interim Report. http://www5.cao.go.jp/keizai-shimon/kaigi/special/future/chuukanseiri/01_e.pdf, accessed on September 25, 2016.

Chino, Masato. (2009). When Is Year One of the Society with Declining Population (Jinko gensho Shakai "Gannen" wa itsuka). *Statistics Today* 9, http://www.stat.go.jp/info/today/009.htm, accessed on September 25, 2016 (in Japanese).

Daiki, Shoki. (2006, January 7). Overcoming the Demographic Challenge (Jinko mondai no kokufuku ni mukete). *Nikkei Shimbun (morning edition).*

Economic Planning Agency. (1992). White Paper on the National Lifestyle -Arrival of a Society with a Small Number of Children: Impact and Response (Kokumin hakusho- shoshika shakai no torai, sono eikyo to taisaku). http://www5.cao.go.jp/seikatsu/whitepaper/h4/wp-pl92-01503.html, accessed on September 25, 2016 (in Japanese).

Employment Measures Act (Koyo taisaku ho). (1966). http://law.e-gov.go.jp/htmldata/S41/S41HO132.html, accessed on September 25, 2016 (in Japanese).

Goto, Hitoshi. (2012). Cabinet Secretariat Reference: Policy on Permanent Resident of Foreign National of Japanese Origin (Nikkei teiju gaikokujin shisaku ni tsuite). http://www.cas.go.jp/jp/seisaku/kyousei/dai1/siryou7.pdf, accessed on September 25, 2016 (in Japanese).

International Labour Organization (ILO). (1952). Social Security (Minimum Standards) Convention (ILO Convention No. 102). http://blue.lim.ilo.org/cariblex/pdfs/ILO_Convention_102.pdf, accessed on September 25, 2016.

Japan Private Sector Trade Union Confederation. (1988). Japan Private Sector Trade Union Confederation's Standards on Prerequisite and Criterion for Accepting Foreign Workers.

Keidanren. (2004). Proposal about the issue of foreign acceptance (Gaikokujin Ukeire Mondai ni Kansuru Teigen), http://www.keidanren.or.jp/japanese/policy/2004/029/, accessed on February 17, 2017 (by Japanese).

Kito, Hiroshi. (2006, May 18). Economics Class Keizai Kyoshitsu: Living with a Declining Population: Prescription for the Future (part 2 of 3) (Jinkogen to ikiru (chu)—Mirai he no shohosen). *Nikkei Shimbun (morning edition).*

Mainichi Shimbun. (1990, March 2). Prime Minister Toshiki Kaifu's Policy Speech Part 3. *Mainichi Shimbun (evening edition).*

Mainichi Shimbun. (2016, February 9) On the Edges of the Argument: Getting Jobs En Masse Began in Pre-War Years (Ron no shuhen: senzen ni sakanoboru shudan shushoku). *Mainichi Shimbun (evening edition).*

Malthus, Thomas. (1798). An Essay on the Principle of Population. London: J. Johnston.

MHLW. (2005). Documents about Declining birth rate (Syoushika mondai tou ni kansuru Shiryou), http://www.mhlw.go.jp/shingi/2005/02/dl/s0201-3d.pdf, accessed on February 17, 2017 (in Japanese).

MIC. (2015, September 20). Japan's Elderly (65 and older) as Seen in Statistics (Tokei kara mita wagakuni no koreisha (65-sai ijo)). *Statistical Topics* No. 90 (Press release), http://www.stat.go.jp/data/topics/pdf/topics90.pdf, accessed on September 25, 2016 (in Japanese).

MOFA. (2001, January 31). Policy Speech By Prime Minister Yoshiro Mori to the 151st Session of the Diet. http://www.mofa.go.jp/announce/announce/2001/1/0131-2.html, accessed on August 9, 2017

MOJ. (1950, May 4). *Public Assistance Act* (Act No. 144). http://www.japaneselawtranslation.go.jp/law/detail_main?vm=&id=24, accessed on September 25, 2016.

MOJ. (2015). Number of foreign resident by nationality/region (kokuseki/chiiki betsu zairyuu gaikokujin no suii), http://www.moj.go.jp/content/001160917.pdf, accessed on February 17, 2017 (in Japanese).

National Diet Library. (1990, March 2). Prime Minister Toshiki Kaifu's Policy Speech, http://kokkai.ndl.go.jp/SENTAKU/sangiin/118/0010/11803020010002a.html, accessed on February 17, 2017.

Nikkei. (2015). 25 Years of Japan's Declining Birthrate (Nippon shōshika no 25-nen dēta to tomoni tadoru), http://www.nikkei.com/article/DGXMZO88001290S5A610C1I00000/, accessed on February 21, 2017 (in Japanese).

Nikkeiren. (2001). The Labor Studies Committee Report (Roudou Mondai Kenkyu Iinkai Houkoku), http://www.jil.go.jp/foreign/event/kouen/20020125/pdf/t3-1.pdf, P17 18, accessed on February 20, 2017 (in Japanese).

Nikkei Veritas. (2014, April 20). Immigration as an Option: Lack of Manpower, Calls for Its Use Suddenly Growing, Treatment as "Convenient Labor," (Imin to iu sentakushi—hitodebusoku, niwaka ni moriagaru katsuyoron, 'Benri na rodoryoku' atsukai, choki dewa kakuho muzukashiku"). *Nikkei Veritas.*

OECD. (2013). International Immigration Outlook 2013, http://static.pulso.cl/20130612/1768132.pdf, accessed on February 17, 2017.

The Council on Population Problems (Ed.). (1974) *Trend in the Population in Japan—Heading for a stationary population.* (Tokyo: The Council on Population Problems)

Tokyo Metropolitan Government. (2013). In Tokyo Metropolitan Government: Types of Special Zones. http://www.seisakukikaku.metro.tokyo.jp/invest_tokyo/english/invest-tokyo/index.html, accessed on September 25, 2016.

Ueki, Yuko. (2011). Review of the Parental Authority System in Order to Prevent Child Abuse—Bill to Amend Parts of the Civil Code, etc. Research Office Committee on Judicial Affairs. (Jido gyakutai boshi no tameno shinkenmondai no minaoshi). *House of Councillors Research Office Document* No.320, 4. http://www.sangiin.go.jp/japanese/annai/chousa/rippou_chousa/backnumber/2011pdf/20110905003.pdf, accessed on September 25, 2016 (in Japanese).

Yamaguchi, Kazuo (2013), *Analytical facts in considering the declining birth rate—Cabinet Office Declining Birth Rate Crisis Breakthrough Taskforce Report, (Shoshika taisaku wo kangaeru ni atatte no bunseki jigyo).* http://www8.cao.go.jp/shoushi/shoushika/meeting/taskforce/k_1/pdf/s7.pdf, accessed on September 25, 2016 (in Japanese).

CHAPTER 6

National Land (Regional/Administrative) Policy

Sunahara Yosuke, Seta Fumihiko, and Umeyama Goro

Japan has been unable to stem the overconcentration of its population in the main metropolitan areas throughout the postwar era, including first and foremost Tokyo. This demographic imbalance has led to the impoverishment of rural villages reliant on agriculture, forestry, and fishing, as well as an overall decline in these industries in the disorderly sprawl of towns in the suburbs of large cities. Sadly, the responses to these risks from overconcentration remain inadequate.

Sunahara. Y
Graduate School of Law, Kobe University, Kobe, Japan

Seta. F
Department of Urban Engineering, Graduate School of Engineering, University of Tokyo, Tokyo, Japan

Umeyama. G (✉)
Consultant, Risk and Crisis Management, Tokyo, Japan

Y. Funabashi (ed.), *Japan's Population Implosion*,
https://doi.org/10.1007/978-981-10-4983-5_6

6.1 "Tokyo Overconcentration" Pronounced Since the Late 1980s

Following World War II, the three major metropolitan areas of Tokyo, Osaka, and Nagoya underwent rapid population growth. Speedy economic recovery and urbanization accompanied the high economic growth that undergirded Japan's revitalization. After leaving junior and senior high school, young people from rural areas moved to metropolitan areas for higher education and employment. However, the regional population did not show much growth. Since 1945, with the exception of a slight recession, people have continued to flow out of the countryside and into the cities. The exodus was mainly fed by the younger generations and this resulted in demographic aging. Although the rural birth rate is comparatively high, it has not led to a large population growth because the overall number of young people remains low. There is a scarcity of young people who can inherit property and businesses in rural farming, forestry, and fishing villages, and depopulation has even gained ground in the regional cities. In this way, there is a steep national lopsidedness between overcrowded metropolitan areas and depopulated regional, especially rural, areas.

Although the population inflow toward urban areas has continued unabated since the end of the war, from the late 1990s, despite a chronically depressed economy, the population influx to Metropolitan Tokyo continues. This differs from the high-growth economic era and the period of stable growth when people were attracted to the various appeals, or *pull factors*, of the three major metropolitan areas centered on Tokyo, Osaka, and Nagoya. These pull factors, such as highly paid jobs, a modern lifestyle, and the wide range of choices and entertainment, were unavailable in the rural regions. During the "Lost Decades" and up to the present, the population influx to Metropolitan Tokyo has been a phenomenon fueled by people motivated by *push factors*, who feel they have no choice but to leave the aging and declining population in rural areas and an accelerated hollowing out of industry in their regional hometowns. Japan's land planning policies have continually attempted to modify the use of national land since the 1960s with the aim of improving income gaps and the disproportionate development caused by this population imbalance and the lack of industrial infrastructure, as shown in data from Ministry of Land, Infrastructure, Transport, and Tourism (MLIT).[1] Examples of this include infrastructure development and incentives to attract industry to regional areas, such as new industrial cities and large-scale projects, as well as the regulation of factory locations in metropolitan areas (Seta 2002).

6.2 Prime Minister Tanaka Kakuei's Sense of Crisis

The intense massing of the population in the big cities during the period of high economic growth exacerbated overcrowding problems in metropolitan areas. Small houses and meager transportation infrastructure sowed the seeds of residential discontent, and pollution from the industrial plants developed into a major health issue in Tokyo and Yokohama.[2] From the mid-1960s, city residents began to express their displeasure with politics. The Liberal Democratic Party (LDP), which had been in power since 1955, was aware that the overcrowding in big cities was a politically important issue. They were worried that if the migration from rural areas to the cities fed by high economic growth continued, it would weaken their rural political base. For example, in his 1963 *The Conservative Party's Vision*, LDP Labor Minister Ishida Hirohide claimed that the interests of new city workers should be guaranteed if rural depopulation and urban expansion was going to destroy the existing base of LDP support (Ishida 1963). More important was the ascent of radical forces in metropolitan areas. After the election of Minobe Ryokichi as Tokyo governor in 1967, candidates supported by left-wing forces (the Socialist and Communist Parties) started winning gubernatorial and mayoral elections in the early 1970s in metropolitan Osaka Prefecture, and the cities of Yokohama, Osaka, Kyoto, Kobe, and Nagoya. Tanaka Kakuei, who served as prime minister in the early 1970s, interpreted the Minobe victory in the Tokyo gubernatorial election as an eruption of frustration by the overcrowding in large cities that was undermining civic life. Tanaka expressed a sense of crisis that the LDP would lose power if it could not respond to this problem (Tanaka 1967). His sense of crisis was reflected in the LDP's *Urban Policy Planning* and the *Plan for Remodeling the Japanese Archipelago*.[3] While addressing the metropolitan overcrowding problem, politicians began to discuss regional development, especially in the rural areas, and "industrial relocation," or the transfer of urban industry to rural regions. To address metropolitan overcrowding, the government passed the "three factory bills"—the Factories Restriction Law, already enacted in 1964, and the 1972 Industrial Relocation Promotion Act/Factory Location Act—making it difficult to build large plants in metropolitan areas. As a result, the national share of industrial shipments fell in metropolitan areas and industrial integration expanded in other regions.[4] However, these changes were not sufficient to cause a reverse population flow from the big cities back to the rural regions.

6.3 Disorderly City Sprawl in Metropolitan Areas

The urban areas continued to sprawl outward during the high-growth period, expanding in a completely disorderly fashion and swollen with inhabitants who had migrated from rural regions. This was due to an inability to implement effective urban planning. National and local governments responded to the influx of residents by constructing New Towns (detailed in the introduction) and the private sector also developed housing estates in the suburbs. Because the private sector carried out its development in makeshift form wherever it could find land for sale, cities began to appear randomly at the edges of urban areas and followed no particular rational plan. Until the 1968 City Planning Act, there was hardly any development regulation. Housing complexes that feed cities require social infrastructure including roads, water supply, or sewer systems, and need to provide a variety of city services such as garbage collection. Both the Japan housing corporations and local governments (prefectural housing supply corporations) generally proceeded with housing and infrastructure development in parallel, but this was not necessarily the case with private developers. Many real estate developers, as soon as they purchased land from farmers, would only construct houses where even the minimum social infrastructure was lacking, then sell these homes to workers in metropolitan areas, and move on quickly to the next project. City expansion advanced as the level of national income rose with Japan's economic growth and increased automobile use. Urban and residential land as well as commercial facilities created new urban sprawl, making it impossible to discern where a town ended.

From early on, local governments viewed this situation as problematic. But in reality, the authorities were unable to impede such growth and they found themselves forced to provide city services even in inconvenient areas far from the city center. This stretched municipal budgets to the breaking point, and local governments on the outskirts of metropolitan areas suddenly hit by waves of development faced similar problems. They implemented various responses aiming to indirectly regulate such indiscriminate development, but with a relatively weak commitment. For example, by adding more in-depth development procedures, they increased the opportunities for local governments to negotiate on undesirable developments. There were even some local governments that created guidelines to curb water supply services and to regulate minimum site areas and residential land allocation.[5]

However, local governments could not do more than that. Their construction administration that controls developments had an especially strong centralized structure and took the stance that regulation not stipulated in national legislation (so-called additions and protrusions) was undesirable. In Shime Town, Fukuoka Prefecture, appeals were made that guidelines to limit city services violated the Constitution, and there—as well as elsewhere—local governments were ultimately forced to provide services. The effect of government-by-guidance was limited. For example, during the high-growth era, some local governments enacted ordinances and guidelines limiting water supplies to newly developed apartment blocks and housing estates over a certain size because they could not provide the commensurate social infrastructure for their rapidly swelling populations. In response, there were many examples of landowners and developers pleading that their right to receive public services on land they were developing had been unduly infringed. There are numerous cases in which the courts have upheld the claims of the landowner and developer plaintiffs. And even when the local government's positions were upheld, the authorities hesitated to introduce further adjustments and regulations constraining future disorderly development for fear of exciting a strong backlash from developers.

Japan's current City Planning Act was enacted in 1968 but the problem of urban sprawl had already gained considerable ground in metropolitan areas. Until the act was passed, city centers were only subject to a certain degree of zoning regulations and regulation of suburban land use was extremely weak. The act urged the deciding of the city planning area and allowed the divison of the area into "urban areas" that could be developed relatively freely, and "urbanization control areas" that curbed development. As a result, the act raised hopes that an aggregated urban area would be formed. In reality, however, many local governments failed to draw such lines and left loose regulation in place. Even if areas were demarcated, one municipality after another designated excessive urban areas in the name of absorbing population growth over the next ten years.

6.4 Apartment "Slumification" Becomes a Reality

Japan's housing policy has centered on home ownership. The goal was to increase the quality of housing, foster the housing industry and stimulate the economy. Subsequently, this policy also led to accelerated aging and depopulation in suburban housing estates in large cities. For the average salaried worker during the high-growth era, the dream was to own a house in the suburbs. The process is clearly explained in *The Modern Housing Game*, created in 1973 by architectural scholar Ueda Atsushi and others. Using an apartment or condominium as a transitory pathway, the detached house is set as the final "winning" destination in this role-playing game. This mirrored housing policy at the time. Furthermore, as previously mentioned in Chapter 2, various problems have emerged regarding apartments because of Japan's declining population. Especially in condominium blocks with a unique system known as unit ownership, residential and social aging is leading to deteriorating living environments. In the long term, it is not an exaggeration to say that it is in condominiums where the greatest social problems reside. Unit ownership was institutionalized in 1962 when condominiums began to proliferate. The system is one where the owner of a condominium has a very strong right of ownership, but entrances and hallways within the building as well as facilities such as elevators and land are essentially jointly owned. There are great temporary advantages to owning and managing facilities jointly. While owners must deal with all problems themselves in a detached house, in condominium blocks management and costs can be shared. As the years pass, however, when important decisions relate to the entire block, such as large-scale repairs and rebuilding, shared ownership can turn into a major obstacle. Even if the advantages are worth the cost, it is not easy to form a consensus among owners who all have different levels of income and different life circumstances (Kawagishi and Yajima 2006).

Three to four decades after construction, owners often change and individual condominiums are rented out. Rebuilding or selling the property requires four-fifths of the owners to move forward and this task is made even more difficult by the fact that it is the corporate body and residents themselves who must negotiate the consensus. Rebuilding apartment blocks will become increasingly difficult in the future. This is because the population decline will lower demand for apartments in city centers. Rebuilding a condominium will become more and more difficult under the rights and obligation of unit ownership as currently legislated. A building subject to unit ownership is a condominium, office building, or one that is separated into two or more rooms, each of which is under a separate unit ownership. According to research by the MLIT, there are only approximately 200 apartment blocks that so far have successfully been rebuilt in the entire country. Many of them have alleviated the owners' costs of rebuilding by increasing floor space and selling the extra portions. In the future, however, selling the increased floor space will not be so easy.

There will be a growing number of cases where rebuilding cannot proceed because owners will have to bear the entire cost of rebuilding on their own.[6]

If apartments are left to deteriorate without rebuilding or repairs, residents will leave to find new housing, leaving behind people who do not have sufficient financial resources or the physical strength to do so. In unit ownership condominiums, the management fees paid by owners and tenants will cover maintenance such as cleaning, elevators, water tanks, or pipes, but if owners disappear or their whereabouts are unknown then management fees fall into arrears and services will grind to a halt. The number of apartments undergoing this kind of "slumification" is unlikely to be low.[7] Housing policies and schemes that have to date aimed at home ownership for the vast majority of the nation and increasing the quality of housing stock no longer fit the reality of Japan's declining population. However, as can be seen from low interest rate loans for home purchase and tax deductions for housing loans, ownership continues to be treated more favorably than renting. This system, premised on high growth, needs to be substantially modified and redesigned to match the changes that Japan is currently undergoing.

6.5 Disorderly Suburban Development Prompted by Subsidies

In contrast to the continually expanding urban areas, what policies have been implemented in regions with no demographic growth? In the 1990s, rural areas in particular increased their reliance on public investment, and government policy placed too much importance on construction, or "building things." Factors behind this included not only the goal of local revitalization but also external pressure for expanding domestic demand due to an increase in Japan's trade surplus and demands for economic measures in the wake of the collapse of the bubble economy. It was difficult for the central government to spend money effectively due to the limited number of these local branches. Instead, the central government encouraged the local governments to expand their public works projects by providing subsidies. These local governments were further expected to act as public works operators and by issuing municipal bonds raised large funds. As a result, the total amount of municipal bonds more than doubled in the period from 1990 to 2000. Sewerage was a typical public works project, which recorded significantly high growth rate in the 1990s, and accounted for more than half the total of local public enterprises bonds in

2010. However, despite laying sewers in anticipation of future expansion, these residential areas failed to grow and often only the negative legacy of debt remained. Maintenance and renewal of infrastructure is difficult because municipalities facing this kind of problem have to repay debt with only a low user base.

Basically, the main sources of revenue for local governments are local taxes from their current residents and municipal bonds, which future residents will pay off. Since these are their own funding sources, local governments are expected to strive for efficient usage. Central regions that offer more efficient investment are to be redeveloped even at the risk of political costs from curbing investment in areas where service provision is costly. Additionally, when the provision of services and improvement of facilities are very costly, local governments are required to move forward with the consolidation of residential areas, to lower service frequency, and reduce costs to a level comparable with areas of higher population density. However, what happens when the central government will definitely repay municipal bonds and the local government does not need to tap into its own resources? In this case it becomes easier to provide services and invest in improving living standards in peripheral areas that missed out on real estate development because these sectors no longer need to focus on efficiency. In fact, government subsidies have been more generous toward low population density areas than central regions and facility improvement. Subsidies for sewerage systems and transportation network development, such as roads, are typical examples. On the other hand, the central government did not welcome the redevelopment investment for central regions.

This trend was prompted by the electoral systems of local councils. Under the single non-transferrable voting system, lawmakers stood a better chance of winning if they strengthened ties with a limited group of supporters. The object was the allegiance of residents who owned farmland and relatively large property holders that had lived in the same area for generations, not people who move repeatedly. It is not easy to obtain a consensus on redevelopment emphasizing transportation infrastructure in central urban areas where various interests clash. Even if it is to raise a region's profile, some residents will be required to relocate. Coordinating these complex interests becomes painful for people living in the area for a long time. Developing areas that lie outside city centers is relatively easy because comparatively these places are lacking in attractive features, such as shopping, parking, infrastructure, or transportation. Gaining agreement on injecting financial resources to make them more attractive is a smoother process. Because land rights are often held by a small number of long-term residents considered "influential," there is also a political benefit for lawmakers to curry their favor. Consequently, farmland is actively diverted into new residential areas resulting in profits for landlords. Aided, of course, by a preference for suburban residences on larger plots, suburbanization has moved forward throughout the Japanese archipelago. When the population grows in these suburbs, it is not hard for leaders such as mayors and deputies to embark enthusiastically on policies to quickly augment amenities for residents, including infrastructure such as roads and sewerage. This, in turn, encourages more votes. On the other hand, it is difficult to embark on reinvestment policies for improving transport access to city centers or providing low-cost, high-quality rental housing, which would raise the efficiency and attractiveness of the city as a whole. Long-term residents hardly gain profits by these projects.

Using government subsidies to improve life in the suburbs also means losing the charm of local city centers. Along with the advance of a society that centers on automobile transport, shops in central areas are unable to compete with the rise of commercial facilities in the suburbs, such as large-scale shopping malls, and they have started to fall into bad financial straits. As a result, central shopping areas have lost their appeal and turned into "shutter streets," where a majority of shops have closed their doors. In turn, this accelerates further urban sprawl as people move out from the center. People who felt unattracted to local cities headed to Tokyo, not to other local cities, in search of a change in lifestyle or more diverse education and work opportunities.

6.6 Unrealized Effects of "Decentralization Reform"

From the 1990s, the state envisioned that if decentralization reform went ahead, cities and towns would become more comfortable to live in and more dynamic through greater efficiency and competition between each region. It is hard to say, however, whether such an ideal was ever realized in many municipalities, and now further population decline spurs forward the demise of a vibrant local community. Decentralization reform, which took off in earnest from the mid-1990s, amended a centralized political system that intervened in regional decision-making, and aimed at local governments determining what was right. In 1999, the Comprehensive Decentralization

Act was passed, abolishing "agency delegation," or local governments mandate delegated from the central government with its supervision. Central and local governments were placed on an equal footing, and rules governing the relationship between the two were overhauled. The amount of power held by the central government were transferred to local government, increasing their discretion.

The next stage focused on decentralizing fiscal authority and building local competence, and each local government then needed to make its own decisions about fiscal and human resources. The problem was how to guarantee fiscal resources for local governments. Having carried out requisite projects via local government, the central government had guaranteed local authorities the necessary fiscal resources in the form of local allocation tax grants and national government disbursements. Fiscal decentralization reduced projects and the accompanying funding resources entrusted by the central government to local governments. The plan was to transfer national taxes into local taxes (fiscal decentralization). The intention was for local governments to be freed from central government delegation, and be allowed to decide whether or not to implement projects on their own. It was thought that this would eliminate unnecessary projects and make local funding of projects more efficient. And it was hoped that local governments would use their money to improve previously developed areas in their region rather than finance new developments.

However, fiscal decentralization did not proceed smoothly because decentralization led to economic disparities between local governments. If funds collected as national taxes were transferred to local taxes, local governments with an adequate local tax base could expect an increase in revenue, but areas with low local taxes could not. In addition, if local allocation tax grants and national government disbursements were cut, it would be a major blow for local governments without an adequate local tax base. Consequently, rich municipalities and other municipalities clashed in their opinions over fiscal decentralization. In 2005 and 2006, when the so-called "Trinity Reform" proceeded under the leadership of the Koizumi administration, the National Governors' Association finally initiated a tax revenue transfer of 3 trillion yen. However, there was a fierce backlash such a policy had actually widened the gap between local governments. Some parties looked for local government to cater to local interests when implementing projects by imparting a degree of fiscal resources and administrative capability. Local government mergers were emphasized as a key measure for this.

Before the so-called Great Heisei Mergers in 1999 there were over 3200 local governments nationwide. More than 2700 of these were towns and villages with a population of 50,000 or less. It was difficult to hire staff with sufficient expertise since municipal offices were also small in sparsely populated municipalities (MIC 2010). Fiscal resources were used inefficiently because economies of scale did not work, and they were also considered vulnerable to the risk of changes in the socio-economic environment. The state asked these small municipalities to merge and expand their scale, so they could act as recipients of a transfer in authority. Between 1999 and 2014, the number of municipalities decreased by more than 1500 and the number of smaller towns and villages fell below a thousand.

Did the local government capability improve after the municipal mergers, and did efficiency also gain ground? Indeed. With multiple municipalities merging, duplicate costs—such as council meetings and public relations—could be reduced. However, even though populations grew through mergers, people were not going to suddenly change where they lived. Pre-merger administrative services provided by local government could not abruptly be terminated—including infrastructure, transportation services, hospitals, and schools—and there was great controversy in areas where some services were eliminated. It is difficult to achieve sufficient efficiency if people remained living dispersed throughout a large area. Municipal mergers were meant to improve the regions, but they caused problems instead. The main cause was the special merger bonds that were to act as a "carrot" to encourage mergers. An incentive was required to assuage municipal representatives whose seats would be reduced, the residents of small towns and villages at risk of being buried by incorporation into a larger government, and to get all to agree to a merger. The special merger bonds comprised debt on more favorable terms than normal municipal bonds. For this reason, many local governments that merged used these special bonds to establish a variety of facilities. However, it is easy for debt to escalate precisely because it easily borrowed. This problem was pointed out from the very beginning, but the impact was unclear given that specific municipal data on the issuance of special bonds was not released. However, recent research points out that although special bonds for post-merger projects replaced normal municipal bonds, which had the effect of reducing the balance, this effect did not surpass the total amount of special bond issues. The outstanding balance of municipal bonds tended to be higher in local governments that merged than in those that did not (Miyashita and Nakazawa 2014).

In summary, the reforms pursued by the government since the 1990s have almost all failed to produce the initially envisioned effect, and have, to the contrary, possibly aggravated the fiscal situation of local governments.

6.7 How to Eliminate the Risk of Overconcentration in Tokyo

With depopulation, regions fall into decline and people leave; however, the overconcentration of people, money, and goods in Tokyo continues. How should this situation be perceived in terms of risk?

An influx of population into metropolitan areas occurs to varying degrees in all countries that have achieved high growth through modernization and industrialization. However, Japan has experienced a quasi-permanent overconcentration in urban areas that persists and continued well into a period of stable growth—this is not the case in all such countries. In fact, there are countries in Europe with small land areas where the population is dispersed. In Asia, the trend toward overconcentration in large cities like capitals continues in developing countries and even in those where high growth has settled down. The concentration of various economic functions in Tokyo expanded in parallel with high economic growth, and became a decisive feature during subsequent globalization. Even in recent years, advanced economic functions remain concentrated in Metropolitan Tokyo (including neighboring prefectures), accounting for 31.9% of the national share of total production in 2008, and with 65.9% of headquarters for corporations with capital of 5 billion yen or more, 84.1% of foreign corporations, and 73.8% of finance (bill clearings), but only 27.5% of the population as of 2009 (Yamaguchi 2011). The decentralization of government ministries and agencies as well as the transfer of authority to local governments is still an on-going process and many important public decisions are still made in Tokyo. The risks of overconcentration are roughly twofold. One is the risk of a physical overconcentration of people and material. Traffic congestion, crowded rush hours, a poorer health environment, and urban pollution are all problems that develop when people amass beyond the capacity of the infrastructure. Moreover, Tokyo faces a very high probability of a major earthquake in the future and there is a strong fear that many people will be affected simultaneously, making measures including evacuation and emergency medical care difficult and escalating the scale of human and economic damage. This could also lead to a catastrophic failure in social infrastructure due to other disasters or terrorism. Another risk is due to the concentration of the systems responsible for national decision-making and information gathering. If these key state functions were impaired in the wake of earthquakes, war, or terrorism, the country as a whole would be paralyzed even though there was no major damage elsewhere. Moreover, because information communication, transportation and distribution networks are all designed around Tokyo, the nationwide impact would inevitably be large if these systems were blocked or compromised. Considering Tokyo's role as one of the world's leading international financial markets, the international fallout would be immeasurable if it should fall into financial or economic dysfunction. This is exactly the kind of a risk that needs to be considered from an international perspective.

With these risks as a backdrop, the term "Tokyo overconcentration" often prompts the negative question of why Tokyo does not decentralize, moving key agencies and hubs elsewhere, or through a process of "multipolarization?" In recent years, economic activities and social infrastructure are moving toward distributed networks and web-like connectivity that are considered safer. Even political decentralization that produces a variety of policies has been recommended. However, if one looks back at the postwar national planning of Japan, overconcentration in Tokyo is not something that can be easily corrected. The background to this problem is that once it was seen to have a positive effect on economic growth. While Japan enjoyed prolonged high economic growth in the second half of the twentieth century and entered the league of advanced nations, Tokyo continued to be at the center of politics, economics, and government administration. Although the threat of a major earthquake was always discussed, keeping disaster risk estimates low and not dispersing the country's central core promoted efficient business activities. The priority was always on economic growth. It is possible to consider this factor as contributing to Japan's economic success. Fortunately, there has been no major disaster in Tokyo since the end of the war, and many believe that the "high-risk, high-return strategy" has paid off.

Talk of transferring the capital's government functions was popular temporarily in the 1980s and 1990s. Those in favor of relocation spoke of curbing soaring land prices and eliminating the adverse effects of overcrowding, promoting decentralization, and alleviating disaster risks such as earthquakes. However, land prices fell as the bubble collapsed, the adverse effects of overcrowding were partially alleviated by the development of infrastructure, and decentralization came to be considered as something different from the physical relocation of government functions. During this period, talk of relocating the capital took a back seat, although the risk of earthquakes was not necessarily reduced. In the Great East Japan Earthquake of 2011, the Fukushima Daiichi Nuclear Power Plant leaked radiation into the atmosphere and exposed Tokyo to radioactive contamination. However, with apologies to the victims, the fact that the nuclear power plant was at a certain distance from Tokyo acted as a major risk-hedge for Tokyo, responsible for decision-making, and for Japan's socio-economy based there. Although the Tokyo Electric Power Company and the state were accused of neglecting and ignoring the dangers of nuclear power, if we look at the distant location of this nuclear power plant from Metropolitan Tokyo, it appears that officials did not necessarily view nuclear power as safe. And that view proved correct.

Nuclear power plants still operate and their reopening is currently being debated in Japan, but that does not mean the risk has gone away. There is fear that once again a nuclear accident will occur and this could throw Tokyo into disarray. Learning the lessons of Fukushima, nuclear power plants should be reopened only after they have adequately demonstrated nuclear safety and security, accident preparedness, training, and the leadership and crisis response capabilities of leaders, including politicians, have been assessed. Moreover, unanticipated crises that may affect Tokyo are not limited to large-scale natural disasters such as earthquakes. The risks of war and terrorism, infectious diseases, a government bond slump, energy crises, and foreign state collapse are also increasing. This is a serious problem for Tokyo as well and requires competence in governance from politicians who are keenly aware of worst-case scenarios as well as the value of new cooperative relationships between the public and private sectors. Recently, new developments in city centers are drawing up business continuity plans and there is cooperation to carry out public–private support for stranded commuters. In residential areas, there are a growing number of examples of government and residents cooperating in looking at ways of evacuation and evacuation training.

The significance of alleviating overcrowding will weaken if population decline progresses even in metropolitan areas, eliminating the overconcentration that has been the focus of a decentralization policy. The side effects and adverse results from deploying economic activity centered in Tokyo will be quite moderate compared to the problem of population growth. The term "Tokyo overconcentration" may even assume a more positive connotation of strengthening the Megalopolis Tokyo to support Japan in global competition. However, disaster preparedness is indispensable if we are to positively judge Tokyo's overconcentration. Of all the major cities of the world, Tokyo's disaster risk is exceedingly high.[8] A wise choice is required for national land use: are unavoidable risks such as earthquakes to be met without dispersing the state's central functions, or is a certain degree of decentralization the answer?

Notes

1. MLIT 1962–2015.
2. See Tokyo 2014; Yokohama 2015.
3. See Tanaka 1972; LDP 1968.
4. According to the *Census of Manufactures* (Ministry of International Trade and Industry 1974), the proportion of the amount of industrial shipment from the major metropolitan areas (Tokyo, Kanagawa, Saitama, Chiba, Aichi, Gifu, Mie, Shizuoka, Osaka, Hyogo, and Wakayama Prefectures) decreased from 66.7% in 1960 to 61.2% in 1974.
5. Local governments began to emerge from the late 1960s in the attempt to restrain haphazard development by establishing residential land development guidelines. The earliest example of this was the guideline that the city of Kawanishi (Hyogo Prefecture), lying in the suburbs of the Osaka metropolitan area, established in 1967.
6. MLIT 2010.
7. Asahi 2013.
8. According to the Munich reinsurance company, risk index of natural disasters in world metropolises (Munich Re 2003), out of the world's 50 major cities, the risk in Tokyo and Yokohama is markedly high.

References

Asahi. (2013, May 17). Increasing uninhabited houses causes slumification (Fueru akiya kanritaisei ga kuzure suramuka ni) (Interview with Kyoji Matsumoto). *Shukan Asahi*. http://dot.asahi.com/wa/2013050800040.html, accessed on September 25, 2016. (in Japanese).

Ishida, Hirohide. (1963, January). *The Conservative Party's Vision* (*Hoshuto no Bijon*). *Chuokoron*, 88–97.

Kawagishi, Umekazu & Hiroaki Yajima. (2006). Study on the process of consensus building on regeneration of apartment complexes (Shugo jutaku no koshin saisei ni okeru goi keisei purosesu ni kansuru kenkyu). Paper presented at the 39th meeting of the College of Industrial Technology, Nihon University,

Architecture panel, December 2. http://www.cit.nihon-u.ac.jp/kouendata/No.39/4_kenchiku/4-093.pdf, accessed on September 25, 2016. (in Japanese).

Liberal Democratic Party Urban Policy Research Council ed., (1968). Urban Policy Planning (Toshi seisaku taiko). LDP Public Relation Council Publication.

Miyashita, Tomohisa & Katsuyoshi Nakazawa. (2014). *An empirical analysis of municipal bonds issued in merged municipalities—Japanese fiscal finances after the 'social security/tax integrated reform'* (*Gappei jichitai ni okeru chihosai hakkenno jissho bunseki*). Tokyo: Yuhikaku Publishing Co. Ltd.

MIC. (2010). Publication on Heisei Mergers (Heisei no Gappei ni tsuite no kouhyou). http://www.soumu.go.jp/gapei/pdf/100311_1.pdf, accessed February 20, 2017. (in Japanese).

MLIT. (1962–2015). Land planning (Kokudo Keikaku) (National land planning reports). http://www.mlit.go.jp/kokudoseisaku/kokudoseisaku_tk3_000026.html, accessed on September 25, 2016. (in Japanese).

MLIT. (2010). Manual of consensus building towards the reconstruction of the apartment (Manshon no tatekaeni kansuru goiteki keisei manyuaru). http://www.mlit.go.jp/common/001064895.pdf, accessed on September 25, 2016. (in Japanese).

Munich Re. (2003). *Annual Review: Natural Catastrophes in 2002*. Munich Re Group.

Seta, Fumihiko. (2002). *Policy for redressing regional disparities and its transformation as the result of globalization* (*Chiiki kakusa zesei seisaku to gurobaruka ni tomonau sono henyo katei*). PhD dissertation, University of Tokyo. http://www.regionalplanning.net/seta/ddd/abstract.htm, accessed on September 25, 2016. (in Japanese).

Tanaka, Kakuei. (1967, June). Reflection on the Liberal Democratic Party (Jiminto no hansei). *Chuo Koron* 82(7), 284–293.

Tanaka, Kakuei. (1972). *A Plan for Remodeling the Japanese Archipelago* (*Nippon retto kaizo ron*). Tokyo: Nikkan Kogyo Shimbun.

Tokyo Metropolitan Government. (2014). Pollution warning and numbers of patients (Kokagaku sumoggu chui hatsurei nissu to kenko higaishasu no suii). http://www.metro.tokyo.jp/INET/CHOUSA/2014/12/60oc9101.htm, accessed on September 25, 2016. (in Japanese).

Yamaguchi, Hirofumi. (2011). Issues in capital functions and a direct earthquake hit on the capital (Shuto chokka jishin to shuto kino o meguru kadai). *National Diet Library ISSUE BRIEF* No.725. http://www.ndl.go.jp/jp/diet/publication/issue/pdf/0725.pdf, accessed February 16, 2017. (in Japanese).

Yokohama City. (2015). Pollution warning and numbers of patients (Kokagaku sumoggu chui ho hatsurei kaisu oyobi todokede higaisha su no keinen suii). http://www.city.yokohama.lg.jp/kankyo/mamoru/kanshi/atrend/106-alm.html, accessed on September 25, 2016. (in Japanese).

CHAPTER 7

The Demographic Issue and Silver Democracy

Kato Sota

Over the last two decades, despite a constant stream of announcements about a looming crisis and a slew of proposed countermeasures, both Japanese politicians and public officials have merely wrung their hands about the country's aging and depopulation problem. The reasons for the long delays in responding to such an important issue are the dearth of "promotional forces" for switching policies, the absence of "social entrepreneurs" with the ability to implement policy, and the issue of "silver democracy," which has been the subject of growing attention in recent years. Nonetheless, it is a mistake to only blame politicians and the government bureaucrats for the population decline. Voters themselves should consider how the future should look and get involved.

7.1 "Too Little, Too Late"

Few policy issues are as easy to forecast as demographic issues. The drop in Japan's birth rate became abundantly clear with the 1.57 Shock of 1990, developing into a social topic of great concern.[1] That same year, the Ministry of Health and Welfare (MHW) indicated in a report

Kato. S (✉)
International University of Japan, Tokyo, Japan

Y. Funabashi (ed.), *Japan's Population Implosion*,
https://doi.org/10.1007/978-981-10-4983-5_7

that the birth rate decline was a "looming demographic crisis."[2] In 1992, the term "low birth rate" was used for the first time in the White Paper on National Life. In response to this shock, academics and think tanks proposed measures for addressing the declining birth rate.[3] Examples of typical measures to reduce the mental and economic burden during the childcare years are the private sector-led provision of nursery services to respond to a wide variety of needs, the establishment of a consultation system to assuage worries about childcare, and special treatment for families caring for children in education, housing, and other related areas. The state also took measures, although small. However, the sense of urgency was minimal and a comprehensive response was delayed.

In a 2005 simulation, the Organization for Economic Co-operation and Development (OECD) showed that a significant recovery in Japan's birth rate to 2.0 was possible (d`Addio and d`Ercole 2005). However, the irony was that the calculations for this estimation also found that the reason there was room for improvement was because of Japan's delayed response to the falling birth rate. In the general election of December 2014, the Liberal Democratic Party (LDP) and the Democratic Party of Japan (DPJ) drafted countermeasures again, but there was an undeniable feeling of being "too late." Even if the birth rate improved dramatically from 2014 onward, it would have taken a considerable amount of time to increase the working-age population (15–65 years of age), and during this interval people's lives and the economy would face serious hardships.

Diverse countermeasures—including those presented in this book—have been examined for their response to Japan's demographic transformations. But now it is necessary that politicians and government officials move boldly in order for drastic measures to be implemented. However, both of these groups have moved too slowly and have arguably even become an obstacle to any significant change. If this remains the case, no matter how excellent the proposed countermeasures are, their success will be in peril unless we first analyze the causes behind this political and bureaucratic delay and find ways to overcome it. Why has the political and administrative response rate been so lackluster? Why was changing Japan's demographic policy continually postponed? In other words, why has it been "too little, too late?" And what is needed to move the issue ahead and to find the necessary countermeasures?

7.2 What Led to a Postponement of the Demographic Issue

Normally, a "policy shift" is required to fundamentally solve any given social problem and these policy shifts need "momentum" for their implementation. This momentum can come from "driving forces," emerging from political parties or politicians, government or the bureaucracy, public opinion, the media, private think tanks, non-profit organizations, and others. Furthermore, often it is also necessary for an influential person to wave the flag for change. Scholars of American politics refer to this flag-waving role as a central driving force for policy change, calling these individuals "political entrepreneurs" or "political sponsors." On the other hand, political resistance due to vested interests can also be expected to oppose such shifts in policy. Typical examples of these "forces of resistance" are labor unions and interest groups such as agricultural cooperatives. Additionally, there are cases when public opinion mobilizes against a given change and itself becomes an opposing force on fiscal issues that involve higher taxes or social security cuts. In sum, whether a policy shift is ultimately implemented or not is determined by the result of the political struggle between the "driving forces" and the "resistance forces." The result is not merely a question of numbers. Strength of cohesion and the presence of political entrepreneurs are also important factors. For example, urban residents and ordinary consumers form the driving force behind many radical policies, often enjoying superiority in numbers. However, because their overall unity is poor, they end up being "a herd of cattle" and are often no match for organized forces of resistance.[4]

In a parliamentary system, it is the Diet members who decide the political and administrative direction. But it is difficult for politicians to become the driving force on policy issues like the demographic problem that require a response over the long term. This is because there are incentives for politicians to postpone long-term issues. There is a maximum four-year term for members of the House of Representatives and six for the House of Councillors. If political parties and members do not appeal to voters with their achievements before the next election, parties may lose power and members may even lose their seats. Thus, the activities of political parties and members prioritize resolving short-term policy issues where short-term outcomes can be obtained, and it is easy for them to be biased toward interest-induced policy that is supported by their constituents. The difficulty for politicians to serve as the driving force on long-term issues applies to democracy in general and is seen as a significant factor in postponing the resolution of such issues.

A typical example of postponing a political problem is Japan's budget deficit, an issue that seemingly plagues all developed democracies. Japan has the largest amount of outstanding public debt in proportion to its Gross Domestic Product (GDP) of any industrialized democracy as the result of massive inflationary packages implemented by the government after the bursting of the economic bubble. A policy shift toward edging up the consumption tax and drastically overhauling the social security system is needed to consolidate public finances in Japan. Despite the many valuable suggestions and estimates put forward by economists, over the decades the problem has been continually postponed. A similar postponement has also taken place with the demographic issue. From a politician's perspective, emphasizing the fact that measures are necessary for the future of society in decades to come would not necessarily translate into votes. In fact, looking at the literature from the late 1980s, when political awareness of the demographic issue first grew into a salient topic, political driving forces to solve this problem remained consistently weak. This occurs not only in the political realm, but also in government circles where the push for solving the demographic issue remained ineffective. In the MHW, major departments responsible for the elderly and pension policy were always given priority while responding to the low birth rate problem was somewhat lackadaisical. The silo mentality that exists between Japanese government agencies and ministerial departments also dampened enthusiasm to draft any proposals to alleviate depopulation. After the enactment of the Basic Act on Countermeasures for a Declining Birth Rate Society in 2003, a minister (Minister of State for Measures for Declining Birth Rate) and department (Council on Measures for a Society with a Decreasing Birth Rate) were set up in the Cabinet Office.[5] However, a negligible number of staff was assigned. Paradoxically, the Cabinet Office's greatest role is in coordination between ministries and agencies, making it impossible for it to play the role of a strong driving force for policy change.

7.3 No Noticeable "Forces of Resistance"

One ironic feature of Japan's demographic decline is that although driving forces are weak there are also few opposition groups. It is as if the problem has been completely ignored. Indeed, several political and administrative factors leading to a postponement of countermeasures for the declining birth rate have been cited in interviews with former policy officials.[6] These include ideological conflicts; conflicts of interest within MHW; the silo mentality over preschools and kindergartens between MHW and the former Ministry of Education, Culture, Sports, Science, and Technology (MEXT); and the expectations of the former Ministry of Finance (MOF), among others. However, while many of these are inherent to any policy implementation or policy change, generally they amount to nothing more than political and administrative barriers. As resistance forces, they are so weak that they are in a different dimension compared to the public backlash over higher taxes and the social security costs of fiscal problems.

So, are there any other significant forces of resistance?

Perceiving the demographic issue as a conflict of "elders vs. the young," it is often said that the elderly are keenly opposed to policy change. Framing this question of the "silver democracy" and its political power is also likely to grow in the future, a concern we will cover in later chapters. To date, however, the simplistic idea that the elderly, who have superiority in numbers in Japan, are using the democratic system to exploit the young has overtaken the actual situation. There are certainly moments when politicians have given priority to measures for the elderly. One example is a case in 2011 where the co-payment rate hike for the elderly under the nursing care insurance system was passed over. MHLW had considered raising the nursing care insurance co-payment rate the previous year, but the DPJ, which was in power at the time, postponed it over fear of the impact on the unified local government elections.[7] But circumstances in which the elderly and their interest groups have attempted to thwart childcare support, for example, have not been observed. Social cracks or conflict between the elderly and the young do not show up in public opinion surveys. Rather, such surveys reveal an altruistic solidarity and community spirit across the generations. It is possible that, to a certain degree, members of the silver democracy have become passive resistance forces.

In terms of the struggle between "driving forces" and "resistance forces" over long-term policy issues, the demographic issue can be imagined as depicted in Fig. 7.1. We find no evidence that driving forces and resistance forces have engaged in a bitter political battle over the demographic issue to date, especially regarding the low birth rate. Rather than having been politically postponed, it is probably more appropriate to view this problem as having been politically and socially ignored.

The question then becomes: why was such a critical issue continually ignored given that no groups were pushing it and few were opposing

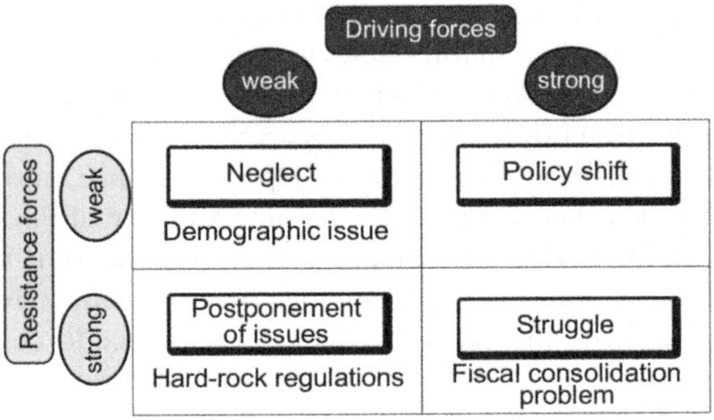

Fig. 7.1 A typology of long-term policy issues: driving forces and resistance forces

change? This is perhaps the most important question to be considered when analyzing factors for Japan's delayed if not absent response to the pending perfect storm of demography.

7.4 An Absence of Strong Political Entrepreneurs

What stands out in Japan's delayed response is the weakness of positive factors, such as a dearth of forces that promote change, rather than negative factors, such as political resistance and political and government incentives to postpone the problem. In this respect, the MOF, which deems a balanced budget to be a kind of ministerial dogma, is a powerful driving force in fiscal issues. Previously there was a powerful group of fiscal lawmakers, the *Okura-zoku*, who had a cooperative relationship with the MOF in the LDP, and there are still many influential proponents of fiscal equilibrium. Even in academia, where fiscal policy research is energetic, the balanced budget faction probably forms the mainstream among leading political economists or macroeconomists. Although there is fierce political resistance to fiscal consolidation, its proponents are also powerful. By contrast, Japan's demographic maelstrom, while equally as important as fiscal issues, and despite being a political issue that requires long-term prescriptions, finds itself with few interested parties and proponents.

As far as can be seen from past literature and interviews with interested parties, MHW and LDP labor lawmakers did not view the population problem as a central policy issue until just before the start of the 2000s, long after the 1.57 shock of 1990 became well-known. Knowledgeable people that were interviewed noted that the majority of scholars and the media within Japan did not initially demonstrate a strong interest in the question.[8] Even when voices were raised about worsening state finances and serious depopulation caused by population decline, falling government bond ratings, and the exodus overseas of top-level companies and young people, as "signs of a lost kingdom," there were no parallel signs that the issue would be tackled in earnest (Komiya 2008). On the other hand, there is also no evidence that MHW or LDP labor lawmakers strongly resisted the implementation of demographic countermeasures. As Fig. 7.1 reveals, in terms of causing a shift in policy the population problem was in a better position because resistance against change was weak. Had strong proponents been abundant, the population problem would have moved to the right column in Fig. 7.1 (i.e., weak resistance forces and strong driving forces), with a high possibility that a shift in policy could have been achieved.

7.5 Former Deputy Chief Cabinet Secretary Furukawa's Commiseration

Iwabuchi Katsuyoshi of the *Sankei Shimbun* pointed out the importance of the declining birth rate from the 1980s and had also formerly served on the government advisory council on population issues. He first mentioned in an interview, the prime minister's failure to take the lead as a factor in the delay in declining birth rate countermeasures.[9] Iwabuchi said that only the prime minister was capable of changing Japanese awareness and the direction of bureaucracy. By contrast, in Germany, Angela Merkel took the lead and declared "Germany will be a child-friendly country in the future."[10] On this point, Ato Makoto, former head of the MHW's Demographic Research Institute, also said "true political leadership" was needed in order to realize a policy shift, stating that senior bureaucrats at MHW and MOF were aware of the seriousness of the problem in the 1990s (Kawai 2014).

A possible political entrepreneur for the population problem was Furukawa Teijiro, a former Deputy Chief Cabinet Secretary. Furukawa was appointed director of the MHW's Children and Family Bureau in 1989, immediately prior to the 1.57 shock announcement. He was one of a few welfare bureaucrats who worked proactively on the demography issue, setting up meetings between 14 ministries and agencies and incorporating for the first time in a government document the commentary on the declining birth rate in the prime minister's policy speech in the same year (Cabinet Office 2004).[11] However, his efforts failed to bear fruit. Citing the reasons for this failure, Furukawa stated that at the time "there was the painful memory of the pre-war 'Beget and Multiply' [slogan] and an atmosphere where the deep-rooted social norm that the public sector should not get involved with births" pervaded. He also noted "the government was flat out dealing with measures for the elderly like the Gold Plan, [a 10-year strategy for the promotion of health and welfare for the elderly] and the zero bedridden strategy."[12] He expressed regret that he had failed to become a political entrepreneur to assist in the resolution of the pending demographic crisis himself, saying about the declining birth rate "I should have staked my life on it" and "I'm still sorry we failed as an administrative agency to come up with results."[13] In short, strong forces of resistance to the population problem were absent. Had at the very least an influential politician, even if not the prime minister, or a leading sector of the administration instead of a lone bureaucrat came out as a driving force, the possibility of progress in shifting policy on this issue could have emerged.

Another way demographic issues differed from fiscal problems was the absence of a consensus on what policies were needed to deal with the problem. There was also an aspect of failing to analyze the cause of the declining birth rate or to identify appropriate actions. Initial measures focused on childcare support. Elsewhere, the analysis of and countermeasures for the critical marriage question, which comprises a series of economic factors, was delayed.[14] American political scientist J.L. Walker has pointed out that the most important role of political entrepreneurs is agenda setting, or establishing the relevant policy issue as an important one (Walker 1974). In sum, there were no influential political entrepreneurs to set such an agenda in the first place, nor were there people promoting large-scale research and analysis—a precondition of agenda setting—and thus it was difficult for either the driving forces or the resistance forces to rally over the population problem.

7.6 Why Did Political Entrepreneurs Fail to Emerge?

Whether or not political entrepreneurs exist for a particular policy issue or not is largely determined by chance. However, what is required is lowering as much as possible this reliance on chance, and a consideration of what kind of environment would be most conducive to the emergence of powerful political entrepreneurs for the population problem. Below we investigate why political entrepreneurs failed to emerge and why environments for fostering them foundered.

7.6.1 Political Entrepreneurs in Politics

A significant factor stymying the voices necessary to champion the next generation of raising children—especially women, who would be the main beneficiaries of implementing new policies as countermeasures for the population problem such as the pursuit of a work–life balance or childcare support—is reflected in the Japanese political system. For example, industrial associations like the Japan Medical Association, the Central Union of Agricultural Co-operatives, and the Japan Federation of Construction Contractors have strongly supported the LDP. The DPJ has mainly received support from organized labor, such as UA ZENSEN, which brings together workers from a wide range of industries, and *Jichiroren*, which consists of local government workers. Organizations incorporated into the major political party machines are organizations such as those of large business owners, large company workers, various industrial organizations, and regional agricultural operators, all of which center on men and the middle-aged. The proportion of female parliamentarians in any developed democracy is also the lowest in Japan.

What is important in making political entrepreneurs stand up for long-term issues is to push key players in the Japanese political process into feeling that the population problem is on their doorstep. They then could be incorporated as a driving force. In this respect the Japan Policy Council, for example, demonstrated the possibility of population decline contributing especially strongly to bringing about an "extinction of local governments" in regional areas (Masuda 2014). This was a beneficial step because it created a looming sense of crisis among members of parliament from the regions, who account for large numbers in the LDP and other parties. It also had a major impact on public opinion. Pushed by this sense of crisis and by the force of public opinion, powerful political entrepreneurs and driving forces may emerge in the future from within the larger framework of regionally elected representatives. Another key aspect of dealing with this issue encompasses the questions of women and the child-raising generation, who have both been inadequately brought into Japan's traditional political decision-making process centered on the LDP. Linking the issue of population decline to elections, which politicians care about the most, would also be beneficial.

An equal weight for each vote is a key principle of democracy. While the Japanese Supreme Court's rulings on the acceptable range for the gap in a vote's weight may become even more severe, they are not likely to be moderated (Kiriyama 2010).[15] "The disparities in the value of one vote" is essentially caused by the fact that sparsely populated areas are allocated an overly generous number of electoral districts. Thus, in order to achieve vote value equality, reviewing the allocation of these electoral districts according to population trends and transferring electoral districts from less populated areas to more populated ones is necessary. If population decline continues, the number of regional electoral districts will also continue to decline on the basis of one-vote equality. Figure 7.2 illustrates our prediction of the number of prefectural seats in 2060 based on population forecasts if electoral districts were redrawn according to one-vote equality. This should make the issue of population decline appear as a much more present danger and hopefully engender a sense of crisis for many Diet members. So what is required to make predictions such as this clear, and, at the same time, create a system for the swift redrawing of electoral boundaries in line with population movements? The transaction costs of having electoral boundaries negotiated each time within and between political parties are far too high. An independent commission on redrawing electoral boundaries should be established, and boundaries redrawn almost automatically in line with changes in population.

Political entrepreneurs are above all required to have an active sense of ownership of an issue. From this point, it would be beneficial in terms of forcing a policy shift for the demographic issue to at the very least increase the proportion of female legislators to the baseline level seen in other developed democracies. Moreover, as industries reorganize and work together to produce new policies—such as promoting a work–life balance through the introduction of shorter working hours and working at home, and childcare support through the provision of a wide variety of childcare facilities—and at the same time makes its voice heard in politics and government, there is a need to nurture politicians to act as their leaders.[16] This, however, is of course accompanied by the downside of leading to a new politics of special interest.

NB of electoral districts	2015	2060	Change from 2015	NB of electoral districts	2015	2060	Change from 2015
Hokkaido	12	11	-1	Shiga	4	4	0
Aomori	4	2	-2	Kyoto	6	6	0
Iwate	4	2	-2	Osaka	19	21	+2
Miyagi	6	5	-1	Hyogo	12	13	+1
Akita	3	2	-1	Nara	4	3	-1
Yamagata	3	2	-1	Wakayama	3	2	-1
Fukushima	5	4	-1	Tottori	2	1	-1
Ibaraki	7	6	-1	Shimane	2	1	-1
Tochigi	5	4	-1	Okayama	5	4	-1
Gunma	5	4	-1	Hiroshima	7	6	-1
Saitama	15	18	+3	Yamaguchi	4	3	-1
Chiba	13	15	+2	Tokushima	2	1	-1
Tokyo	25	37	+12	Kagawa	3	2	-1
Kanagawa	18	25	+7	Ehime	4	3	-1
Niigata	6	5	-1	Kochi	2	1	-1
Toyama	3	2	-1	Fukuoka	11	12	+1
Ishikawa	3	3	0	Saga	2	2	0
Fukui	2	2	0	Nagasaki	4	3	-1
Yamanashi	2	2	0	Kumamoto	5	4	-1
Nagano	5	4	-1	Oita	3	3	0
Gifu	5	4	-1	Miyazaki	3	2	-1
Shizuoka	8	8	0	Kagoshima	5	3	-2
Aichi	15	20	+5	Okinawa	4	4	0
Mie	5	4	-1	Total	295	295	0

Prefectures with reduced seats	30
Prefectures with increased seats	8

(Notes)

- The number of electoral districts is for the single-seat lower house.
- The populations of each prefecture in 2060 are estimated with an assumption that the change rate between 2035 and 2040 will continue from 2040 to 2060, and then adjusted to meet the total population projection for 2060.
- The number of voters of each prefecture in 2060 is estimated with the eligble voters-to-population ratio in 2040.

Fig. 7.2 Number of parliamentary seats in 2060 if the value of votes are equalized (forecast). Source: IPSS, Japan's Regional Population Projections, March 2013

7.6.2 Political Entrepreneurs in Government Administration

There was a strong possibility that government bureaucrats would be more proactive as policy entrepreneurs and take a more long-term approach to the population problem. Bureaucrats, with their superior organizational and research skills, are a keen necessity to face medium- to long-term challenges such as demographic issues, as well as for other issues that require policy planning based on meticulous research.

Since the 1990s, when the population decline became apparent, the theory of an overly simplified "political leadership" was emphasized, and the view circulated that the bureaucracy need only follow the orders emanating from politicians. From this perspective, government administrations cannot be blamed for standing still on the population problem. Excessive political power produced irresponsible bureaucrats lacking a sense of ownership. However, the political parties and the government administration have different constitutional positions and organizations, political neutrality being required of the government administration. An appropriate measure of role-sharing between the government and the public sector is necessary. Political parties are in charge of legislation and do their work in the Diet, while administrative officials are in charge of administration and do their work in the ministries and other administrative agencies that are under the control of the cabinet, being responsible for, among other things, developing plans to implement policies that the Diet determines.

Population decline is an issue that involves a wide range of policy areas, and therefore one authority, such as the MHW, cannot realize policy changes because they reach across issues and confront the Japanese government's silo mentality. For example, the former MOF was probably reluctant to craft countermeasures for the falling birth rate because it believed the elderly could get by without relying on their children, if only social security could be strengthened. There were other barriers created by the silo mentality within the ministries. In the MHW, the key departments that wielded power were the bureaus for pensions and health care, and thus aging was given priority over the falling birth rate. The latter issue eventually came to be handled by the Cabinet Office, which is a coordinating organization as described above and possesses neither the delegated authority nor the organizational culture for drafting policy, so it was impossible for it to become a driving force. A ministerial post on the falling birth rate was subsequently established and staff assigned, but the state seemed to be more interested in the PR impact than content; for example, appointing a woman as the minister as an image-building strategy. Under the DPJ Administration, which only lasted for three years and three months, as many as ten people were appointed as a sort of a revolving door, a situation hardly conducive to a steady approach.[17] There were other barriers between government ministries as well as between their respective agencies. For example, because aging was given priority over the falling birth rate as a policy issue, the MHW was a key department with power over pensions and health care. To break down these types of barriers, the impetus must come from the political side. At the same time, ministries and their agencies could serve as political entrepreneurs, supporting powerful politicians by showing them detailed data and simulations as well as by setting political agenda, from a long-term point of view.

Encouraging political entrepreneurs from the government probably requires structural changes. They need to focus constantly on the population issue and become a powerful driving force, just as the MOF acts on financial issues. In this regard, various efforts have been made, including the establishment of the Ministerial Conference on the Promotion of Measures to Cope with Fewer Number of Children and the enactment of the Basic Act for Measures to Cope with Society with Declining Birth Rate in 1999. However, taking a step further and establishing a "Ministry for the Next Generation" should also be considered. A unified approach is hardly likely because policy related to children consist of a patchwork of measures across multiple government agencies, such as the MEXT and the MHLW. If there were one "ministry" in charge, then it would be easier to conduct more advantageous budget negotiations with the MOF as well.

To date, bureaucrats prepared to "stake their life on the problem," as Furukawa put it, have failed to position the population problem in the proper light. This is evident, for example, from the inadequate number of staff assigned to the Cabinet Office department set up after the enactment of the Basic Act for Measures to Cope with Society with Declining Birth Rate described above. In addition, excessive political leadership to the extent that denies government officials from owning the issue should be reconsidered.

7.6.3 *Other Political Entrepreneurs: Voters, Researchers, Media, and Think Tanks*

Ultimately, it is votes and public opinion that move politics. Compared to politicians who face a continual cycle of elections, voters can take a more long-term view on policy. If voters took a long-term view and chose political leaders that share this kind of thinking, the perspectives held by politicians can have the flexibility to better consider the future.

There are two major challenges hindering the ability of voters to take a long-term outlook on the population problem and becoming the driving force for a shift in policy. The first challenge is the amount of information that voters have about a given problem. Compared to politicians and bureaucrats, voters have less access to information on the population problem. Without data to visualize the future, voters cannot afford the time to analyze the issues themselves and make policy decisions, and even if such data were in the public domain, the public is busy with their own jobs so they rely on their representatives. In order for these voters to make appropriate decisions from a long-term perspective, politicians and government administrators need to present them with future demographic prospects in an easy-to-understand format. In short, the population problem needs to be "made visible." However, not all data on the population problem are made public. Moreover, on the government side, there was a tendency to have an unreasonable hope for a third baby boom and to underestimate the problem. Data showing a clear relationship between the birth rate and female labor participation rates and childcare support were never made available to voters. It is necessary for the government to present all the data and the results of its analysis in an easily comprehensible format to voters if they are to be asked to take a long-term view and make prudent judgments. Furthermore, analytical methods should not just consist of hoping for a future solution (like a baby boom), but should rely on meticulously collected data gleaned from a variety of sources. In addition, the roles of research institutes, the mass media, and think tanks are crucial and can act as driving forces linking the government and voters. Researchers and think tanks examine government data and the media conveys and deciphers it to voters. Indeed, this book is also one recent attempt to do so.

7.7 Will the Silver Democracy Create an Intergenerational Conflict?

The second problem hindering voters from becoming a driving force behind a shift in policy is the possibility of intergenerational conflict, or the possibility of dominance by a "silver democracy."

The population problem is often seen as a still-decades-away issue, and, on the one hand, many elderly voters believe they will not be alive when the actual problem becomes serious. In this manner they give top priority to their own interests and become supporters of measures for themselves, and may also even unite against future policy shifts to alleviate the population problem. On the other hand, the next generation, who by definition should be urging a policy shift for the population problem, cannot become a driving force because they remain powerless in democratic elections at the moment.

Older voters, who form Japan's majority, will exploit the young and the next generation. This is the problem of a silver democracy. It is currently the subject of close attention in many developed democracies that are suffering a drop in population, and the imbalance poses a fundamental question about whether democracy can fully function while facing a steep population decline.[18] A silver democracy is often spoken of in the context of fiscal and social security issues. This is because, with the elderly wielding power in the democratic process, resource allocation by the state is directed toward measures for the elderly—such as augmenting social security budgets—at the expense of those trying to combat a declining birth rate with childcare measures.

The possibility of a silver democracy is especially noteworthy in Japan. It has been pointed out that Japan will be the first nation to enter the age of silver democracy because, compared to other advanced democracies, aging is occurring at a faster pace and its budget deficit to GDP ratio is the largest (Yashiro 2016). Some economists propose introducing a system of generational constituencies in order to solve the problem of a silver democracy, which will have ramifications for the very foundations of the democratic system (Ibori and Doi 1998; Takeuchi 2013). The scale of Japan's current budget deficit most likely cannot be explained without referring to its rapidly advancing demographic aging. Furthermore, democratic institutions have overcome a mountain of criticism and weathered crises without undergoing major institutional reform. Is silver democracy the only issue it cannot overcome? This point needs to be considered carefully and calmly and in light of past wisdom.

Compared to the elderly, Japan's youth are becoming a "minority." The median voter theorem is a basic concept in political science. The theorem states that a majority rule voting system will select the outcome most preferred by the median voter. In terms of age, the median voter in Japan as a whole is currently aged 53 years (Yashiro et al. 2012). But there are also many regional areas where the median voter age is higher than 60, and these constituencies have a greater number of seats allocated to them. Japan leads the world with this structure of an "elder majority, young minority," but it will spread throughout developed democracies and the world in the future.

7.8 "Middle Group" Crucial for Preventing the "Tyranny of the Majority"

It is shortsighted, however, to conclude there the elderly will exploit the youth of Japan because the question of intergenerational conflict is not intrinsically different from the conventional problem of "the majority vs. the minority." And to date democracy has somehow overcome in many cases the issue of "the tyranny of the majority."[19] For example, intellectuals at the time of the French Revolution, such as John Reeves (1752–1829), the barrister who established the Association for the Preservation of Liberty and Property against Republicans and Levellers, feared recrimination and the plundering of wealth of the propertied class by the proletariat. However, at least in developed democracies, such plundering of wealth has to date not been a serious problem.[20] No country to date has introduced the "electoral division by income." Thomas Piketty's book *Capital in the Twenty-First Century*, which has attracted much attention, uses detailed data to illustrate the current situation of wealth accumulating in the top 1%, but this concentration of wealth occurs even in developed democracies.[21] Incidentally, Piketty notes that when the number of children per family falls, resulting in a decline in the number of overall children, it is easier for estates to be inherited intact without any distribution, making it likely that the wealth disparity will be maintained (Piketty 2014).[22]

Various mechanisms have been proposed throughout history for preventing "the tyranny of the majority." James Madison, the fourth president of the USA, proposed two measures in *The Federalist Papers* No. 10, from the famous collection of essays arguing for the ratification of the US Constitution by the state legislatures. The first is to deny direct democracy or to adopt a system of representative government. Madison argued that the intelligence and sense of justice of elected government representatives would alleviate the problem of the tyranny of the majority because they would serve as a buffer from the general public. He pointed out that it would be possible to choose better representatives in a larger electorate since the pool of candidates would also be larger. The second measure was the realization of what is now referred to as pluralism. If various interests compete in a broad electorate, Madison pointed out that the majority and minority groups representing a given interest would stand out and the possibility of the majority persecuting or exploiting minority would be minimized (Hamilton et al. 1788).

Applying Madison's observations to the generational question for modern Japan would run as follows. Regarding representative government, it is not the role of lawmakers to respond only to the will of the elderly because there are many different groups in their constituencies. Rather, based on their intelligence and sense of justice, it is expected that their political behavior will also take into consideration the future of the youth of Japan—the next generation and those not yet born. In reference to pluralism, there are a variety of interests in Japan; for example, workers versus capitalists, men versus women, particular religious believers versus non-believers, and consumption tax proponents versus opponents. When these various groups intersect, projecting "the elderly vs. the young" as the only "social cleavage" is incorrect and the exploitation of the young minority by the old majority will not necessarily happen.

Like Madison, the nineteenth-century French political thinker Alexis de Tocqueville was also concerned about the tyranny of the majority, and he emphasized the importance of a "middle group" for democracy to overcome it and function properly (de Tocqueville 2003 [1835]). American political scientists Robert Putnam, Robert Leonardi, and Raffaella Nanetti named de Tocqueville's middle group "social capital" and argued that it had a critically important role for a well-functioning democracy (Putnam et al. 1992). Compared with other majority–minority problems, this question of a middle group is mitigated in the elderly–youth problem. The potential cleavage between these two groups is assuaged because their relationship is based on the same and most basic unit of human society, the family.[23]

In fact, few empirical studies carried out to date by political scientists have identified a conflict in awareness leading to a social rift between the elderly and the youth. Rather, it is more common to find a strong altruistic awareness between them. These empirical results differ greatly from the assumptions of some economists, but a major factor is that the family functions as a quasi-middle group spanning the generations. In this way—at least to date—the use or awareness of the potential use of democratic institutions to exploit the young and the next generation by the elderly has generally not been observed. An altruistic awareness between generations is strong compared to other majority–minority issues. However, politicians, always looking for more votes, have actually given priority to employment measures, benefits, and others for the elderly.

In that negative sense, there is a possibility that in the balance between driving forces and resistance forces for the population problem, silver democracy has leaned toward the latter.

7.9 Need for "Cooperative Business with Future Generations"

In terms of the population problem, what kinds of responses are necessary for a silver democracy to function effectively for all of society?

As we have already seen, a large part of the intergenerational question can be boiled down to a normal majority–minority question. Furthermore, in this case, unlike other such issues, aspects of intergenerational altruistic and cooperative awareness can be seen between the majority and the minority. These are the elements of the so-called *social capital*. Given their fortuitous existence, placing too much emphasis on the problem of silver democracy has the downside of fanning conflict between the generations and actually damaging social capital. Instead it is important to accumulate social capital, and to maintain and expand the altruistic awareness that has been nurtured to date between generations through the resurrection of a conservative awareness that the British thinker Edmund Burke called an intergenerational contract.

Therefore, it would be more appropriate for the time being to deal with Japan's silver democracy not with any dramatic measures such as generational electorates, but by proceeding with responses that have been used to date to solve other problems in democracy: specifically, a rapid response to the question of redrawing electoral boundaries and making the population problem more visible through the disclosure of information to voters. Of course, the problems a silver democracy poses are still possible in theory. The impact will need to be carefully assessed and, if necessary, some form of response made in the future. However, problems can be greatly alleviated through such "ordinary" measures as reapportioning parliamentary seats. What is required is to first move forward with the obvious matters necessary for a healthy democracy to function.

7.10 Expectation of a "Third Baby Boom" Backfire

There are other factors reluctant to the postponing or masking of Japan's population problem. Situational awareness and political values, as well as institutional problems, became barriers to any potential policy change. Policies in all countries are the by-product of the combined awareness and values of the political system, voters, and policy authorities (Kato 2014). As such, when making policy changes, it is not possible to fully respond under the existing political system, and public consciousness and values often also lag behind the status quo. From this perspective, neither the "family" faction, who think parenting is the responsibility of the parent, nor the "home" faction, who believe pregnancy and childbirth are solely matters of individual choice, consider Japan's population problem as one in which the state should intervene. With such a deep-rooted belief among the people, the government's response was delayed. A point specific to the population problem that can be cited among the various factors that hampered a shift in policy was the miscalculation of future demographic projections. Many policy makers and demography scholars predicted a "Third Baby Boom," hoping it would solve the population problem. When the Third Baby Boom failed to occur, responses were significantly delayed. Thus, releasing various types of government data and analysis by think tanks and scholars is required.

Notes

1. Cabinet Office 2015.
2. MHW 1990.
3. See, for example, Economic Planning Agency 1992; IPSS 1993; Cabinet Office 1998.
4. See Olson 1965. For example, where the objective of a group ends up not being achieved when there are a large number of members because it becomes difficult to create or maintain the group with the objective of securing political and economic benefits.

5. Cabinet Office 2004.
6. Interview conducted by author, personal communication, May 14, 2014.
7. Nikkei 2012.
8. Makoto Ato, personal communication, May 14, 2014.
9. Iwabuchi, Katsuyoshi, personal communication, May 14, 2014.
10. Ibid.
11. The Liaison Council for Ministries and Agencies concerning the Development of an Environment in Which Children Are Born and Nurtured Healthily was established under the Cabinet Secretariat in 1990. On the basis of the "Compilation" published by the Council, the Child Care Leave Act was enacted and support was provided for balancing work and childcare. These measures were later expanded as the "Angel Plan."
12. The Gold Plan, officially the Ten-Year Strategy for Promoting Health and Welfare of the Elderly, was enacted against the background of the rapid progression of aging. It was aimed at developing infrastructure and human resources for in-home welfare as well as developing public facilities for nursing care (IPSS 2014; MHW 1989).
13. Furukawa, Teijiro, personal communication, January 15, 2015.
14. The Basic Orientation of Future Measures to Support Childcare (called the Angel Plan) established in 1994 highlighted support for mothers who had babies and their families. Priority measures under this plan were support for balancing work and childcare (e.g., widespread establishment of workplace crèches) and nurseries that match the various needs of users (e.g., more facilities that provide care after 6 p.m. and facilities that accept zero to two-year-olds) (MHW 1994).
15. For a while after the end of World War II, the Supreme Court's determinations regarding the "disparities in the value of one vote" were very lenient. Since 1976, when the Supreme Court first issued a judgment on this matter, judgments were issued that declared unconstitutionality or "state of unconstitutionality" using a four-to-one disparity as a rule of thumb. This meant that judgment was issued in 1986 that declared a 2.92-to-1 disparity as "constitutional." However, from 2009 on, the Supreme Court has gradually become more strict, as stricter decisions have been coming down, such as a judgment declaring that a larger than twofold disparity is a "state of unconstitutionality." This trend can also be seen in high court judgments as well.
16. Cabinet Office 2013.
17. Fukushima, Mizuho, Hirano, Hirofumi, Genba, Kōichirō, Okazaki, Tomiko, Yosano, Kaoru, Renhō, Okada, Katsuya, Nakagawa, Masaharu, Komiyama, Yoko, Nakatsuka, Ikko.
18. Ferguson 2012.
19. Of course, this fear is tragically embodied in the tyranny of the National Convention during the French Revolution and the rise of Nazi Party under

the Weimar Constitution. Post-revolution confusion in France led to the revival of the Empire, and the Nazi Party experience saw the introduction of "fighting democracy" in Germany.

20. This appeared in its most prominent form in the proletariat revolutions of non-democratic nations.
21. *Capital in the Twenty-First Century* was first published in French in 2013, then in English and Japanese translations in the next year.
22. Piketty also stated in a symposium held in Japan (hosted by Asahi Shimbun and the French Embassy on January 29, 2015) that a declining birthrate exacerbates disparity and there is a need for long-term population growth.
23. Since families are to be found in every society, they are not usually purposely classified into middle groups and social capital, but their impact on democracy is much larger than normal middle groups.

References

d'Addio, Anna Cristina & Marco Mira d'Ercole. (2005). Trends and Determinants of Fertility Rates in OECD Countries: The Role of Policies. *OECD Social, Employment and Migration Working Paper* No.27 (November). http://www.oecd.org/els/family/35304751.pdf, accessed on February 27, 2017.

Cabinet Office, Government of Japan. (1998). To build a society in which homes can be created and children can be reared with hopes for the future (Proposals) (Yumearu katei zukuriya kosodategadekiru shakai wo kizukutameni (Teian)). (White paper, Council of Experts Considering Responses to the Declining Birth Rate). http://www.kantei.go.jp/jp/singi/syoshika/981228teigen.html, accessed on September 25, 2016. (in Japanese).

Cabinet Office, Government of Japan. (2004). White paper on measures for a society with a decreasing birth rate FY2004 (Heisei 16 nenban shoshika taisaku hakusho).. http://www8.cao.go.jp/shoushi/shoushika/whitepaper/measures/w-2004/pdf_h/honpen.html, accessed on September 25, 2016. (in Japanese).

Cabinet Office, Government of Japan. (2013). A 2013 declining birthrate white paper.. http://www8.cao.go.jp/shoushi/shoushika/whitepaper/measures/english/w-2013/index.html, accessed on September 25, 2016.

Cabinet Office, Government of Japan. (2015). A 2015 declining birthrate white paper.. http://www8.cao.go.jp/shoushi/shoushika/whitepaper/measures/english/w-2015/index.html, accessed on September 25, 2016.

Economic Planning Agency. (1992). Whitepaper on national lifestyle (Kokumin seikatsu hakusho).. http://www5.cao.go.jp/seikatsu/whitepaper/h4/wp-pl92-000h1.html, accessed on September 25, 2016. (in Japanese).

Ferguson, M.L. (2012). *Sharing Democracy.* Oxford: Oxford University Press.

Hamilton, Alexander, John Jay, & James Madison. (1788). "The same subject continued (The union as a safeguard against domestic faction and insurrection)." *The Federalist Papers,* No.10. http://www.gutenberg.org/

files/1404/1404-h/1404-h.htm - link2H_4_0010, accessed on September 25, 2016.

Ibori, Toshihiro & Shiro Doi. (1998). *An economic analysis of Japanese politics.* Tokyo: Bokutaku-sha.

IPSS. (1993). Report (Hokokusho).. http://www.ipss.go.jp/publication/j/shiryou/no.13/data/shiryou/syakaifukushi/473.pdf, accessed on September 25, 2016. (in Japanese).

IPSS. (2013). Regional Population Projection for Japan, 2010–2040.. http://www.ipss.go.jp/pp-shicyoson/e/shicyoson13/t-page.asp, accessed on September 25, 2016.

IPSS. (2014). Social security in Japan 2014. http://www.ipss.go.jp/s-info/e/ssj2014/005.html, accessed on October 11, 2016.

Kato, S. (2014). Valley of institutional change: Japanese political economy 1990–2005. *US Midwest Political Science Association (MPSA) Papers.*

Kawai, Ran. (2014). Interview with Ato Makoto: Want to know more about declining birthrates (Yappari shiritai shoshika no hanashi).. http://igs-kankan.com/article/2014/05/000893/, accessed on September 25, 2016. (in Japanese).

Kiriyama, Keiichi. (2010, May). Disparities in the value of one vote enter an era of a twofold disparity as the unconstitutionality threshold (Ippyo no kakusa, iken-rain wa nibaijidai e). *Sekai No. 804.*

Komiya, Ryutaro. (2008, August 14). Economic lecture (keizai kyoshitsu). *Nikkei Shimbun (morning edition).*

Masuda, Hiroya. (2014). *Extinction of the regions: rapid decrease in population caused by population concentration in Tokyo (Chiho shometsu: Tokyo ikkyoku shuchu ga maneku jinko kyugen).* Tokyo: Chuo Koron Shinsha.

MHW. (1989). Ten-year strategy for promoting health and welfare of the elderly. *Japan social security documents IV (1980–2000) (Koreisha hoken fukushi suishin jukkanen senryaku).* http://www.ipss.go.jp/publication/j/shiryou/no.13/data/shiryou/souron/17.pdf, accessed on September 25, 2016 (in Japanese).

MHW. (1990). Hearing about future of Japan's family and childcare (Korekarano kazoku to kosodate ni kansuru kondankai hokokusho). http://www.ipss.go.jp/publication/j/shiryou/no.13/data/shiryou/syakaifukushi/396.pdf, accessed on September 25, 2016. (in Japanese).

MHW. (1994). Basic orientation of future measures to support childcare (Kongo no kosodateshien no tameno shisaku no kihonteki hoko ni tsuite). http://www.mhlw.go.jp/bunya/kodomo/angelplan.html, accessed on September 25, 2016 (in Japanese).

Nikkei. (2012, November 26). Will policy prioritizing the elderly change? *Nikkei Shimbun.*

Olson, Mancur. (1965). *The logic of collective action: public goods and the theory of groups.* Cambridge MA: Harvard University Press.

Piketty, Thomas. (2014). *Capital in the twenty-first century*. Cambridge MA: The Belknap Press of Harvard University Press.

Putnam, Robert D., Robert Leonardi, & Raffaella Nanetti. (1992). *Making democracy work: Civic traditions in modern Italy*. Princeton: Princeton University Press.

Takeuchi, Kan. (2013). Who should be the focus?: the elderly or future generations? *Chuokoron* 128(4), 120–123.

de Tocqueville, Alexis. (2003 [1835]). Jerald E. Bevan, tr. *Democracy in America and two essays on America*. London: Penguin Books.

Walker, J.L. (1974, September). Performance gaps, policy research, and political entrepreneurs: Toward a theory of agenda setting. *Policy Studies Journal* 3(1), 112–116.

Yashiro, Naohiro. (2016). Cut a swathe into "silver democracy" by cutting down on social insurance expenditures instead of raising taxes (Zozei no kawarini shakaihoshohi sakugen dankode "shiruba minshushugi ni kirikome"). *Diamond Online*, June 1. http://diamond.jp/articles/-/92233, accessed on September 25, 2016. (in Japanese).

Yashiro, Naohiro, Manabu Shimasawa, & Nao Toyoda. (2012 July). Overcoming inter-generational conflict through the social security system: beyond silver democracy, *NIRA monograph series* No.34. http://nira.or.jp/pdf/monograph34.pdf, accessed on February 20, 2017. (in Japanese).

CHAPTER 8

Population Growth Policy and Public Finance

Oguro Kazumasa

When the Liberal Democratic Party–Komeito coalition came back to power in December 2012 under Prime Minister Abe Shinzo, it launched a three-pronged economic booster plan: "bold monetary easing, flexible fiscal policy, and a growth strategy that encourages private sector investment." These quickly became known as the "three arrows" of *Abenomics*. With market expectations weakening the yen and lifting the stock market, a sense pervaded in the media that the Japanese economy would be painlessly revived if a 2% annual inflation rate could be achieved. At the end of 2014, Prime Minister Abe also noted at the Council on Economic and Fiscal Policy that "we can reduce the ratio of accumulated debt by enlarging the gross domestic product."

However, now this belief is being revealed as an illusion, the limits of monetary policy are exposed through the fact that the yen is no longer weak and the stock market out of steam. Fiscal consolidation cannot be achieved by economic growth alone; a combination of economic growth, tax increase, and expenditure cuts is essential. The "Economic and Fiscal Projections for Medium to Long Term Analysis" that the Cabinet Office made public in January 2017 shows that the primary balance in Fiscal Year (FY) 2020 will be left with a deficit of 1.4% of the gross domestic product

Oguro. K (✉)
Faculty of Economics, Hosei University, Tokyo, Japan

Y. Funabashi (ed.), *Japan's Population Implosion*,
https://doi.org/10.1007/978-981-10-4983-5_8

(GDP) even under the premise of both a high growth rate and a two percentage point consumption tax hike to 10% in October 2019.[1]

8.1 Three Ideas for Reducing Social Security Costs

When considering social security reform, it is necessary to keep in mind that both beneficiaries and taxpayers are citizens. The more generous the social security is, the greater the people's burden. In Japan, where social security costs continue to expand, we have already entered a bizarre situation that social security, which should guarantee the right to life, is instead threatening the survival of the nation as a whole. When Japan was growing rapidly, it was possible to "tailor the clothes to the body." But now that the economy has matured and growth is not to easily come by, it is necessary to "tailor the body to the clothes." Those clothes, moreover, must be durable—sustainable enough to cope with future population decline—in the form of systems that can be adapted to a declining population going forward.

To this end, it is necessary to change Japan's economy using the following three ideas.

1. *Shift benefits that are skewed to the elderly and redirect them to all generations.* Under Japan's current social security system, a very large proportion of financial resources are devoted to pensions, elderly health care, and long-term care. To reduce this, resources should be directed toward rectifying the declining birth rate on par with France and Sweden, which both have high birth rates.
2. *Change the income redistribution function of social security.* As mentioned in Chapter 4, the majority of financial resources for existing social security are insurance premiums. Since premiums are determined in proportion to income, the function of income redistribution is limited in regards to strengthening this portion of financial resources. If, however, the increase of the consumption tax can expand public funds, it would be less difficult to provide generous allocations to those in the low-income strata of society. Therefore, while maintaining a system where premiums form the mainstay of financial resources, benefits according to income should be enhanced by utilizing public funds. A conscious effort should be taken to skew

the allocation of financial resources toward the child-rearing generation in the aim of helping to reverse the declining birth rate.

3. *Move forward with selecting and concentrating the content of social security*. In the case of social security, it is often thought that raising benefits is good and lowering them is bad. However, if benefits increase, so will the people's burden. The system should be streamlined and prioritized so that benefits are delivered efficiently and effectively to those who need them.

For social security, the quality is important, but it is even more important to focus on public confidence in the system. Since lives cannot be lived over again, it is inevitable that social chaos occurs when people, who have created a long-term life plan premised on a given system, are suddenly faced with a change in that system. Pensions are a case in point, but the same also applies to the declining birth rate. Because raising children also takes many years, parenting choices cannot be made with confidence when considering the costs involved unless measures are reliable. Once a solid system has been created, it is also important not to fiddle with policy. In pursuing social security and fiscal reform, political consensus against a backdrop of a sense of national crisis is also indispensable. For the moment, however, there seems to be no growing sense of crisis within Japan. This is related to the fact that the government and the ruling parties have not released long-term estimates on national finances. Analysis and discussion of medium- and long-term social security costs does not develop because key numbers are not available, making the crisis difficult to see. Behind this absence lies the reality that if long-term estimates are released, political leaders will be hemmed in between spending cuts and tax increases if it will become apparent that if social security spending is not curtailed, public finances will not stabilize unless the consumption tax is raised to over 30%.

However, many countries publish their long-term estimates on public finances. For example, the European Commission publishes a "Fiscal Sustainability Report."[2] It also draws up an "Ageing Report" every three years, estimating and publishing social security costs (pensions, medical and nursing care, etc.) to GDP up until 2060.[3] In the USA, the Congressional Budget Office forecasts the financial situation for 75 years in the future in its "Long-Term Budget Outlook 2014."[4] Based on these predictions, it has published baseline and alternate scenarios to explore the direction of spending cuts and tax increases. In the UK, the Treasury annually publishes the fiscal outlook for the next 30 years, which is supplemented by a long-term financial report for the next 50 years or so.[5] In these countries, people have information on hand concerning how much tax increases and spending cuts are needed to raise the sustainability of public finances, and are able to make judgments about policy direction. The reason that the Japanese government and the ruling parties will not make long-term fiscal estimates public is that the authorities—first and foremost, the Cabinet Office and the Ministry of Finance (MoF)—are reluctant to do so in deference to the political actors, who in turn fear that the media and the public will overreact. Politicians will be at a loss if they are confronted with the all-too-harsh reality of the public coffers. This will invite a backlash from the political side, and the bureaucrats could face punishment through their future assignments. Besides, even if the government tried to make long-term fiscal estimates public, it is likely that the ruling parties would withhold their consent.

8.2 Nine Scenarios Stemming from the Falling Birth Rate

When considering the medium- to long-term shape of the macro economy and public finances in an ongoing situation of population decline, declining birth-rate countermeasures and immigration policy are important factors in addition to fiscal and social security reform. According to data from the National Institute of Population and Social Security Research, in 2013 the average number of children a woman will bear (the total fertility rate, or TFR) in Japan is 1.43, well below the 2.07 needed to maintain the population.[6] The IPSS has predicted that the population will fall by half by 2083.[7] Facing such a crisis, drastic countermeasures are required for the declining birth rate that brings the number of children people have closer in line with the ideal number and reduces the unmarried rate that continues to climb sharply. In 2010, the ideal number of children a couple cited was 2.4. If this was achieved by implementing drastic measures, such as doubling the childcare support budget to 1% of GDP, the birth rate would recover to about 1.6, and the halving of the population be postponed until 2102. Additionally, the birth rate would be further increased if the unmarried rates for men and women in their early thirties (47.3% and 34.5% in 2010, respectively) were reduced.

What impact would drastic countermeasures to the declining birth rate have on the macro economy and finances? Let us consider the following nine scenarios based on a simulation model (Oguro and Takahata 2013).

Scenario 1:	Consumption tax raised to 10% in 2017. No increase in childcare support or reduction in pension benefits.
Scenario 2:	Childcare support budget doubled from 2020, financed by the consumption tax.
Scenario 3:	Childcare support budget doubled from 2020, financed by wage taxes.
Scenario 4:	Childcare support budget doubled from 2020, financed by capital taxation.
Scenario 5:	Childcare support budget doubled from 2020, financed by public debt.
Scenario 6:	Half of pension insurance premiums covered by the consumption tax from 2020.
Scenario 7:	Ten percent reduction in pension benefits from 2020.
Scenario 8:	Consumption tax raised to 15% in 2020.
Scenario 9:	Consumption tax raised to 15% in 2020. Childcare support budget doubled.

The simulation results, summarized as changes in the population and TFR, are provided in Figs. 8.1 and 8.2. In Fig. 8.1, the dotted line

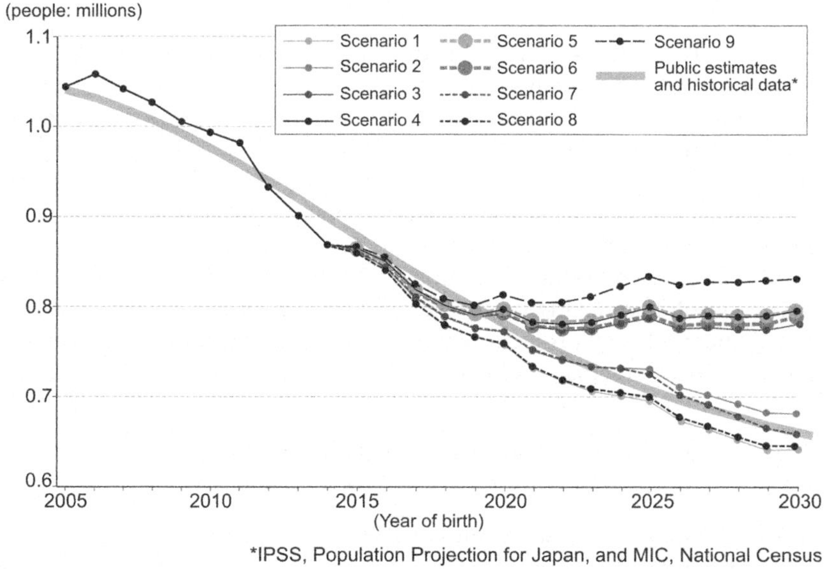

Fig. 8.1 Change in population by year of birth. Source: Compiled from Oguro and Takahata, Child benefits and macroeconomic simulation analysis, 2013

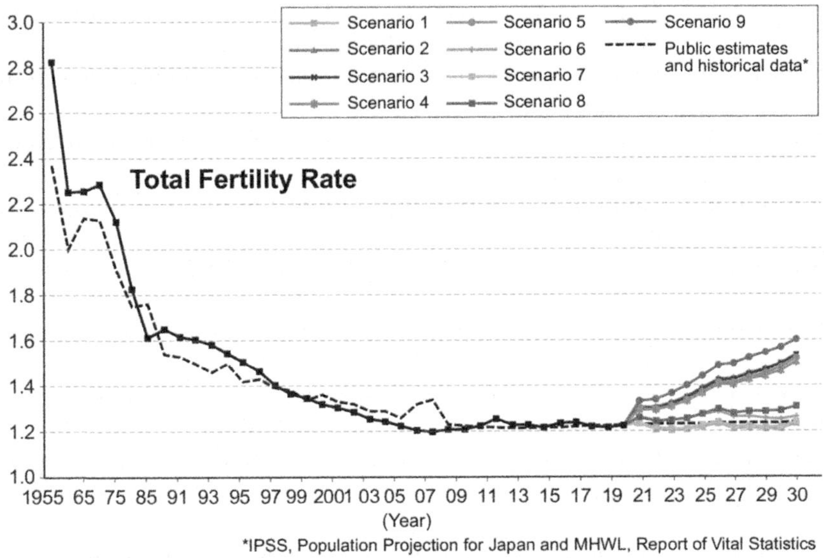

Fig. 8.2 Change in the birth rate. Source: Compiled from Oguro and Takahata, Child benefits and macroeconomic simulation analysis, 2013

represents the 2012 estimate (median births and deaths) of the IPSS. Changes in this dotted line and the "reference case" (Scenario 1) are almost identical. On the other hand, in the "childcare support expansion cases" (Scenarios 2–5) and the "fiscal consolidation cases" (Scenarios 8 and 9), we see a population increase in those born after 2020 as compared to the reference case. Elsewhere, generational population declines in the "pension reform case without childcare support expansion" (Scenarios 6 and 7). The generational population of 2030 is the highest in Scenario 9 (fiscal consolidation plus doubling childcare support). In Fig. 8.2, the dotted lines represent projections from past data and IPSS's 2012 estimates (birth and death median). The TFR of 2030 is the highest in Scenario 9.

8.3 Low Family-Related Spending in Japan Compared to European Countries

Initiatives taken in economically advanced European countries serve as excellent reference points regarding what measures can be taken for a declining birth rate. France and Sweden, with their high birth rates, indicate that not only scale but also balance is important in the financial allocations for childcare support with measures that are unbiased to either the cash benefits (such as family allowance) or benefits in kind (such as childcare services). A report by the OECD also proposes a balanced three-way policy mix of cash and tax, childcare leave, and childcare.[8] The scale of financial resources allocated at such a time should aim at 3% or more of countries with high birth rates like France, the UK, and Sweden, bearing in mind that Japan's family-related spending (family allowance, maternity/childcare leave, other subsidies and benefits in kind) remained around 1.35% of GDP in 2011.[9] Countries with a high ratio of family-related spending promote distinctive policies with respect to child rearing. Sweden has enhanced childcare leave and childcare services with an emphasis on a work–life balance. France pays a family allowance from the second child on. It also provides a broad range of incentives for larger households, including tax incentives, housing, public transport, and recreation. In other EU countries as well—not just these two—the importance of "benefits in kind" such as childcare has been pointed out.

Japan, on the other hand, has a poor record of providing for both cash benefits and benefits in kind. This is the result of attention being drawn to the immediate issue of aging. Well-rounded policies from social security to employment policy seen in other countries have not been deployed. Given that the level of benefits in kind is especially low, parental support measures such as childcare need to be further enhanced.[10] This is consistent with the needs of the child-rearing generation and is also rational in terms of rectifying unmet needs for the birth rate.

8.4 The Social Security Switch from the Elder to All Generations

A clear gap remains between high birth-rate countries and Japan in the "intergenerational balance" of how much social security benefits are spent on which generation. Compared to elder-related benefits, the proportion of family-related expenditure is relatively high in high birth-rate countries as seen in Fig. 8.3. Although there is no great difference between Sweden and Japan in the ratio of pensions and medical care to national income, given that the overall scale of social spending (FY 2009 GDP ratio) is higher in Sweden (30.24%) than Japan (22.58%), the share of other benefits including family-related spending is also relatively high. For example, France now spends more than Sweden on social security

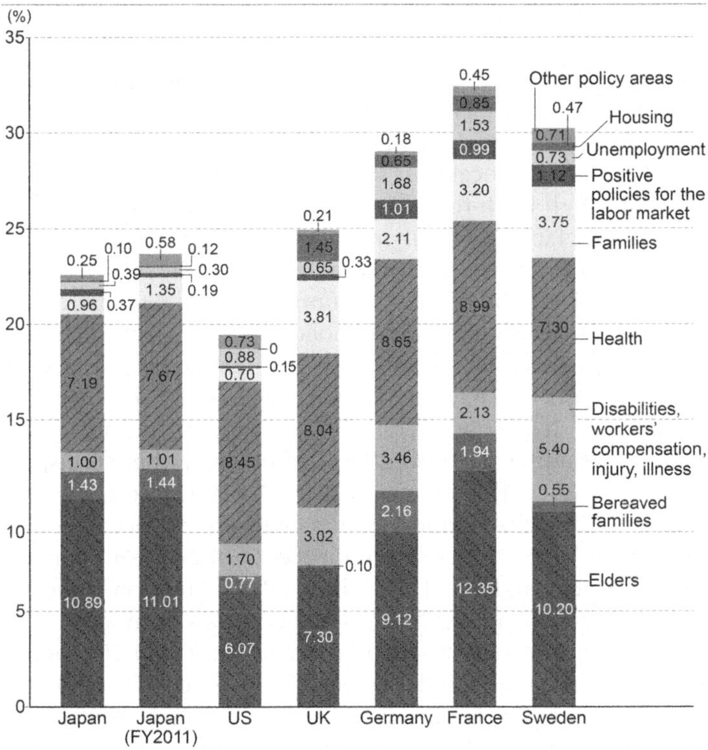

International comparison of social expenditure (FY2009)

Social expenditure	Japan	Japan (FY2011)	US	UK	Germany	France	Sweden
Social expenditure to GDP	22.58%	23.67%	19.45%	24.91%	29.00%	32.41%	30.24%
(Reference) to National income	31.07%	32.31%	24.10%	31.76%	37.98%	43.37%	42.68%

Total fertility rate (2009)

Japan	Japan (FY2011)	US	UK	Germany	France	Sweden
1.37	1.39	2.01	1.94	1.36	1.99	1.94

Fig. 8.3 International comparison of social expenditure to GDP. Source: OECD, Social Expenditure Database, 2013a; Cabinet Office, National Accounts of Japan (for Japan), 2013b; OECD, National Accounts, 2013b (for other countries); IPSS, Population Statistics, 2014

(32.41%). Although its spending on pensions (12.35%) and health expenditure (8.99%) are higher in comparison to Japan, the difference in family-related benefits are even more striking.

This gap is the result of developed countries having switched the focus of social security from the elderly to all generations through a process of selection and concentration, rather than evenly expending social security benefits. In fact, looking at changes in the proportion of family-related expenditure as a percentage of social spending between FY 2001 and FY 2009, it rose from 9.88% to 12.39% in Sweden and has remained high in France from 9.86% to 9.88%. Correspondingly, other social expenditure has been made more efficient, curbed, or cut. In contrast, although this proportion has risen from 3.46% to 4.27% in Japan, its share in total social expenditure is low compared to other developed countries. Consequently, as seen in data from Ministry of Health, Labour and Welfare (MHLW), a situation has developed where the number of children waiting for preschool care has increased to 23,167 children in April 2015, and the quality of childcare has not been improved.[11] Additionally, this low share suggests that any attempt to implement a bold switch to an all-generation type of social security would require a major overhaul of funding allocations in social security as a whole and, correspondingly, substantial pain. As an OECD report points out, given that both the ideal number of children for a Japanese couple and the rate of singles hoping to marry are high (86.3% of males and 89.4% of females in 2010), the possibility of a recovery in the birth rate through enhanced measures is likely.[12] However, speed is of the essence. Going on past policy, these changes will take a considerable amount of time to achieve. Child- and family-related expenditure as a percentage of social security benefits, which was 5.6% in fiscal 1975, dropped through 1998 and 1999 to 3.3% even though the birth rate was already falling, and finally rose again to 5.2% in 2010.

A typical example of why a quick response is necessary can be seen in the elimination of children waiting for preschool care. Establishing nursery centers will awaken latent demand, but that does not mean that the number of children seeking to be accepted would be infinite. Potential demand for preschool care simply needs to be calculated from around 80% women working, as is the case in Sweden and France. In fact, in the Japanese government's 2010 plan, "Children and Childcare Vision," the target for authorized preschools has been set at 2.41 million places in anticipation of potential demand.[13] The fact that France, despite repeated changes in government, has worked consistently throughout the postwar era to enhance family policy is thought-provoking (Shimizu 2007). Public confidence cannot be gained if policy implementation takes time and the specifics frequently change. Premises of effective policy include not only the scale of financial resources and a sense of speed, but also policy stability, continuity, and reliability. Furthermore, the relationship between demographic policy and the questions of poverty and inequality have not yet been fully discussed. The fact that the relative poverty rate has been worsening in recent years, especially the relative poverty rate of children in single-mother households that stands at a very high level in Japan compared to other countries, means that this environment is not one in which people can give birth and raise children without significant difficulty. Problems such as non-permanent jobs, NEETs (people Not in Education, Employment or Training), and social recluses suggests there is a lack of government attention being paid to the employment and living environments of the young, who in turn then hesitate to marry.

8.5 "Four Challenges" Holding the Key to Birth-Rate Recovery

The key to Japan's birth-rate recovery lies in marriage, pregnancy, and childbirth mostly for the strata of society between 20 and 30 years old. Late marriage has been the major cause of the falling birth rate in recent years, mainly because today's young Japanese find themselves in a situation where their future prospects are uncertain. Marriage and childbirth are, of course, a personal lifestyle choice, but they are also influenced by exogenous factors such as employment and the economy, as well as social security. In that respect, it is noteworthy that the postwar baby boom was a common phenomenon throughout world. A major reason was the optimistic hopes for the future. Policies need to support social needs and be provided throughout an individual's lifecycle, beginning with social security in the first half of life. Maintaining stability with no drastic policy swings is also important in terms of creating trust. In response to these obstacles, the following four changes are currently required in Japan.

1. Challenging historical taboos (values)
   The population problem is related to individual values. The harsh memory of the pre-war "Beget and Multiply" slogan remains. However, a consensus must be built on a future population policy. The same applies to the foreign workers' issue, where the government has failed to show a clear direction.

2. Challenging administrative barriers (silo mentality)
   The population problem involves a wide range of administrative bodies. The relevant organizations need the organizational competence to integrate policy across government agencies and to conduct policy in a timely and appropriate manner. A comprehensive approach in terms of social security, not just among border immigration authorities, is especially necessary to advance the foreign workers' issue.
3. Challenging time lags (timing)
   The population problem is like a creeping, lifestyle-related disease. It takes time to resolve, but if left untreated the disease will increasingly erode the host. Therefore, policy speed is required. Action is necessary to avoid the repeat of Japan's so-called lost decades.
4. Challenging the rationality barrier
   Not getting married or having children, which seems rational in the short term and in terms of the individual, also affects the long-term state of the nation. It can lead to a "synthesis of errors," where the results of rational actions by individuals combine to produce results that are not rational for the society as a whole. The difficulty of the population problem is how to reach a compromise. From the perspective of building a sustainable society, support is needed in a way that helps realize people's hopes in line with their lifestyles, while respecting individual choice.

There are a variety of barriers to a resolution of the population problem and these are the most difficult elements. Countermeasures for the declining birth rate in particular will not necessarily have an immediate visible effect. This can be even seen in France, which has invested significant financial and other resources over the last several decades. If nothing is done, however, Japan's decline will continue unabated. Whether policy can be far-sighted and gain a national consensus as well as be boldly decisive is the key to the success or failure of population measures.

8.6 The Crux Is Japanese Workers' Compensation

While foreign workers were also discussed in Chapter 5, here we consider the impact of this problem on the macro economy. It is often thought in general that expanding the intake of foreign workers will have a negative impact on the economy. This fallacy is believed because it is thought that while accepting skilled people will bring certain benefits, the acceptance of unskilled people will lead to a decline in wages for Japanese workers, creating economic disadvantage. This is a deep-rooted myth that misses the corporate point of view and differs from the macro point of view.

According to an analysis of the labor migration effects from developing to developed countries, even if wages fall in developed countries by accepting a greater number of unskilled foreign workers, the increase this brings in terms of future corporate profits is greater than the amount of wage decrease (Grubel 1994). When the government properly distributes an increased portion of corporate profits, workers in developed countries should become economically prosperous. In other words, a greater intake even of unskilled foreign workers may contribute to the Japanese economy. The question then becomes how to compensate Japanese workers disadvantaged by an uptick of foreign workers. Responses taken in other countries are helpful when considering this.

One method is to additionally tax companies that take advantage of foreign unskilled labor, but not to exceed the increase in profit obtained from this utilization, and to redistribute it to the domestic workforce. For example, Singapore applies a Foreign Worker Levy on companies that hire foreigners, which differs according to industry and skilled/unskilled categories.[14] Another method is to additionally tax unskilled foreign labor. The first motivation for unskilled foreign workers to seek employment in other countries is likely the prospect of earning higher wages than in their home country. So, within a range that does not exceed the benefits they obtain (i.e., the wage difference with work in the home country), unskilled foreign workers are taxed additionally and this is redistributed to home workers. Singapore, among others, also takes this approach. It can work in countries that accept foreign workers while implementing this kind of compensation. However, accepting foreign workers does not necessarily mean a drop in domestic worker wages. When foreign workers take the role of supplementing them, it is also conceivable that worker wages do not fall (Oguro 2010). In fact, there are some empirical analyses that show how foreign workers can raise the wages of lesser skilled Japanese workers, who are thought to be able to compete easily with them (Nakamura et al. 2009).

8.7 Connecting Internationalization to Growth

A related issue raised by an expanded intake of foreign labor is that of "social integration," meaning to what extent foreigners, who prolong their stay and bring their families over, are provided with rights such as education and social security and are incorporated into Japanese society. Foreign workers also receive a pension when they retire. Therefore, socially integrating these groups has disadvantages in the long term of squeezing public finances and social security. On the other hand, there are also some benefits from the social integration of foreign workers. For example, because they also pay social insurance premiums, the public pension burden on home workers grows lighter. Tax revenue and consumption also expand. As the number of people supporting pension and medical care decrease in an aging society, it can be expected that young foreigners would supplement this.

These advantages and disadvantages are perceived differently by each generation. We calculated the generational effect for 2015 onwards. To do so, we used "utility" (defined as satisfaction obtained from lifetime wages, or the pension added to take-home wages after subtracting tax and insurance premiums) and consumption in the following three scenarios: (1) an annual intake of 150,000 immigrants; (2) an annual intake of 75,000; and (3) no intake at all (Shimamasa and Oguro 2010). The results found that in cases (1) and (2), utility increased for the generation born after 1980 (Fig. 8.4). This demonstrates that the advantages of a permanent intake of foreigners outweighs the disadvantages and has a positive effect on finances and growth.

What tends to be forgotten is that Japan has set a course of internationalization with its *Cool Japan* growth strategy, attracting foreign tourists and accepting exchange students. There were 13.4 million foreign tourists and 184,000 exchange students in 2014.[15] Discussions on foreign labor assume that if necessary the numbers will remain high, but this will not be easy if the aim is to attract high-quality labor such as those who are highly skilled professionals.[16] Developing acceptance infrastructure is also important in the social defensive sense to prevent the deterioration of public security. But we also need systems that attract quality labor overall. We do not wish to build a house of cards in terms of employing non-Japanese without viewing foreign workers as fellow residents and developing a livable social infrastructure

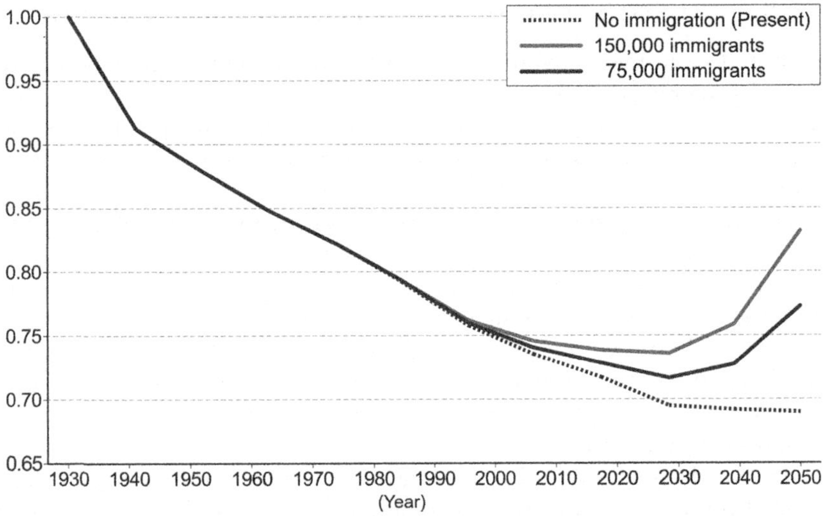

Fig. 8.4 Immigration intake and its generational utility. Source: Compiled from Shimamasa and Oguro, Impact of immigration on the Japanese economy, 2010

for them. Exchange students are the most realistic targets to ensure highly skilled foreign workers. According to the OECD's *International Migration Outlook 2014*, 15–30% of international students remain in their country of study for marriage or work.[17] Looking at foreign students in Japan, 67.0% remained in Japan, of these 24.7% gained employment and 27.1% chose further education.[18] However, the long-term retention rate is not high and many challenges remain, including the unsatisfactory work environment.

One of the significant structural changes that Japan's economy currently faces is the rapid competition from emerging regions, including Asian countries. For Japan, where aging and population decline has progressed and the scope for domestic demand to expand is limited, the expansion of markets in the Asian region can be a great opportunity to realize growth. But if Japan fails to capture this market and misses out on sharing the benefits of this expansion in emerging markets, its position in the world economy, which has retreated in the past two decades, could plunge even further. Capturing the Asian market has huge implications for the future of the Japanese economy. Because of Japan's demographic problems, its finances and economy are definitely heading toward a contracted balance. Consequently, Japan as a whole is doing its utmost to survive and can no longer afford to think about the future from a long-term perspective. Today, now that its GDP, long the second largest in the world, was in 2010 overtaken by China, there is the fear that the Japanese economy will continue to stagnate over the medium to long term if no action is taken.

To prevent Japan's future malaise, reform through a revival of both a growth strategy for public finances and social security is required, as is utilizing foreign workers. This has the potential to contribute to robust changes in the future. Accepting immigrants would encourage the resurgence in domestic demand for Japanese companies as well as fan various related demands. Japan appears to be at precisely the moment when it should proceed with bold reform, including the utilization of foreign workers, with a view to the interest of its future generations.

Notes

1. Cabinet Office 2017.
2. European Commission 2016.
3. European Commission 2015.
4. Congressional Budget Office 2014.
5. UK Office for Budget Responsibility 2015.
6. IPSS 2015.
7. IPSS 2012.
8. OECD 2014a.
9. Cabinet Office 2014.
10. Cabinet Office 2008.
11. MHLW 2015.
12. Cabinet Office 2013a; OECD 2014a.
13. Cabinet Office 2010.
14. Details of the levy can be found in the Ministry of Manpower, Singapore Government website. Retrieved from: http://www.mom.gov.sg/passes-and-permits/work-permit-for-foreign-worker/foreign-worker-levy
15. National Tourism Organization 2014.
16. The concept of "highly skilled professionals" is explained in Immigration Bureau of Japan (2015).
17. OECD 2014b.
18. Student Services Organization 2015.

References

Cabinet Office, Government of Japan. (2008). Depopulation and social policy 2008 (Heisei 20 nen shoshika shakai taisaku hakusho) (White paper). http://www8.cao.go.jp/shoushi/shoushika/whitepaper/measures/w-2008/20webhonpen/index.html, accessed on September 23, 2016 (in Japanese).

Cabinet Office, Government of Japan. (2010). Vision of Children and Young People (White paper). http://www8.cao.go.jp/youth/suisin/pdf/vision_english.pdf, accessed on September 23, 2016 (in Japanese).

Cabinet Office, Government of Japan. (2013a). A 2013 Declining Birthrate White Paper (Heisei 25 nen ban, shoshika shakai taisaku hakusho). http://www8.cao.go.jp/shoushi/shoushika/whitepaper/measures/english/w-2013/, accessed on September 23, 2016.

Cabinet Office, Government of Japan. (2013b). National Economic Count Report 2013 (heisei 25 nen ban Kokumin Keizai Keisan Nenpou). http://www.esri.cao.go.jp/jp/sna/data/data_list/kakuhou/files/files_kakuhou.html, accessed on February 23, 2017 (in Japanese).

Cabinet Office, Government of Japan. (2014). A 2014 Declining Birthrate White Paper (Heisei 26 nen shoshika shakai taisaku hakusho). http://www8.cao.go.jp/shoushi/shoushika/whitepaper/measures/english/w-2014/, accessed on September 23, 2016.

Cabinet Office, Government of Japan. (2017). 2017 Council on economic and fiscal policy (Heisei 29 nen keizai zaisei shimon kaigi). http://www5.cao.go.jp/keizai3/econome/h29chuuchouki1.pdf, accessed on January 25, 2017 (in Japanese).

Congressional Budget Office. (2014). Long-Term Budget Outlook 2014. https://www.cbo.gov/sites/default/files/113th-congress-2013-2014/reports/45471-Long-TermBudgetOutlook_7-29.pdf, accessed on September 23, 2016.

European Commission. (2015). *The 2015 Ageing Report: Economic and budgetary projections for the 28 EU Member States (2013–2060)*. Luxembourg: Publications Office of the European Union. http://ec.europa.eu/economy_finance/publications/european_economy/2015/pdf/ee3_en.pdf, accessed on September 23, 2016.

European Commission. (2016). *Fiscal Sustainability Report 2015* (Institutional Paper 18). Luxembourg: Publications Office of the European Union. https://ec.europa.eu/info/publications/fiscal-sustainability-report-2015_en, accessed on September 23, 2016.

Grubel, Herbert. (1994). The economics of international labor and capital flows. In Giersch, H. (Ed.), *Economic aspects of international migration*. Berlin: Springer-Verlag.

Immigration Bureau of Japan. (2015). Points-based System for Highly Skilled Foreign Professionals. http://www.immi-moj.go.jp/newimmiact_3/en/system/index.html, accessed on September 23, 2016.

IPSS. (2012). Population statistics in Japan (Nihon no shorai jinko). http://www.ipss.go.jp/syoushika/tohkei/newest04/gh2401.pdf, accessed on September 23, 2016 (in Japanese).

IPSS. (2014). Documents about demographic statistics (Jinkou Toukei Shiryousyu). http://www.ipss.go.jp/syoushika/tohkei/Popular/Popular2014.asp?chap=0, accessed on February 23, 2017 (in Japanese).

IPSS. (2015). *Population statistics data book in 2015 (Jinko tokei shiryoshu)*. Tokyo: National Institute of Population and Social Security Research. http://www.ipss.go.jp/syoushika/tohkei/Popular/P_Detail2015.asp?fname=T04-03.htm, accessed on October 13, 2016 (in Japanese).

MHLW. (2015). Research about the kindergarten and crèche (Hoikujo kanren jokyo matome) (Press release). http://www.mhlw.go.jp/file/04-Houdouhappyou-11907000-Koyoukintoujidoukateikyoku-Hoikuka/0000098603.pdf, accessed on September 23, 2016 (in Japanese).

Nakamura, Jiro, et al. (2009). *Japan's foreign labor force* (*Nihon no gaikokujin rodoryoku*). Tokyo: Nikkei Publishing (in Japanese).

National Tourism Organization. (2014). *Marketing Data in 2014: Japan*. Tokyo: Japan National Tourism Organization.

OECD. (2013a). OECD Social Expenditure Database 2013. http://www.oecd.org/els/social/expenditure, accessed on February 23, 2017.

OECD. (2013b). National Accounts 2013. http://www.oecd-ilibrary.org/economics/national-accounts-at-a-glance-2013_na_glance-2013-en, accessed on February 23, 2017.

OECD. (2014a). Japan: Advancing the third arrow for a resilient economy and inclusive growth ("Better Policies" Series, April). https://www.oecd.org/japan/2014.04_JAPAN_EN.pdf, accessed on September 23, 2016.

OECD. (2014b). International Migration Outlook 2014. OECD Publishing. http://ekke.gr/ocd/wp-content/uploads/2014/12/SOPEMI-2014-E.pdf, accessed on September 23, 2016.

Oguro, Kazumasa. (2010, September 8). Foreign immigrants—positive impact on macro-economics (Gaikokukara no jinzai ukeire—makuro ekonomikusu niwa purasu). *Nihon Keizai Shimbun*. http://www.rieti.go.jp/jp/papers/contribution/oguro/01.html, accessed on September 23, 2016 (in Japanese).

Oguro, Kazumasa & Junichiro Takahata. (2013). Child benefits and macroeconomics simulation analyses: An overlapping-generations model with endogenous fertility, *Public Policy Review*, 9(4), 633–659.

Shimamasa, Manabu & Kazumasa Oguro. (2010). Impact of immigration on the Japanese economy: A multi-country simulation model. *Journal of the Japanese and International Economies*, 24(4), 586–602.

Shimizu, Yasuyuki. (2007). Family policy in France (Furansu no kazoku seisaku). *Overseas Social Policy Research (Kaigai shakai hosho kenkyu)* No. 161, 50–60.

http://www.ipss.go.jp/syoushika/bunken/data/pdf/18624306.pdf, accessed on September 23, 2016 (in Japanese).

Student Services Organization. (2015). Report on number of foreign students in Japan, fiscal year 2014 (Heisei 26 nendo gaikokujin ryugakusei zaisekijokyo chosakekka). http://www.jasso.go.jp/about/statistics/intl_student_e/2014/__icsFiles/afieldfile/2015/10/20/data14.pdf, accessed on September 23, 2016 (in Japanese).

UK Office for Budget Responsibility. (2015). Fiscal sustainability report. http://budgetresponsibility.org.uk/docs/dlm_uploads/49753_OBR-Fiscal-Report-Web-Accessible.pdf, accessed on September 23, 2016.

CHAPTER 9

Optimum Population/Land Relocation: Area Management—Crisis Management

Sunahara Yosuke, Seta Fumihiko, and Umeyama Goro

The population problem is closely related to the question of how many people are living in a given location. As seen in Chapter 6, Japan currently has been unable to control unregulated development and this imbalanced dispersal of the population has led simultaneously to advancing regional decline, caused by overconcentration in Tokyo, and a severe degradation of the suburbs, not only in large cities but also in small- and medium-sized ones. What can be done to improve the situation? We describe below two strategies that should be employed together, and, at the same time, show several ways of dealing with the national risk that overconcentration in Tokyo causes.

Sunahara. Y
Graduate School of Law, Kobe University,
Kobe, Hyogo, Japan

Seta. F
Department of Urban Engineering, Graduate School of Engineering,
University of Tokyo, Tokyo, Japan

Umeyama. G (✉)
Consultant, Risk and Crisis Management, Tokyo, Japan

Y. Funabashi (ed.), *Japan's Population Implosion*,
https://doi.org/10.1007/978-981-10-4983-5_9

9.1 The Long-Term Perspectives to Realize the "Compact City"

Raise the density of regional cities; maintain area city services

The first strategy is to revive the regions and raise the population density of local cities. On the one hand, the higher the population density is, the greater the ability to cover a larger number of residents with fewer resources. Thus infrastructure (roads, bridges, water and sewage, etc.), public transport (buses, trains), public facilities (public halls, elder facilities), and public services (elder care and welfare, etc.), which are becoming increasingly difficult to coordinate, could be operated and managed more efficiently. On the other hand, as the population continues to fall, there is no choice but to relocate residents to designated areas. It is not enough to enhance a region merely by building new public facilities such as libraries in the heart of a city or trying to attract commercial facilities such as department stores; it is necessary to use innovative approaches like narrowing residential areas and turning them into safe, convenient, and comfortable places that attract people.

So, what specifically needs to be done? The recent hot topic of "compact cities" (Fig. 9.1) is one idea that offers an effective way to increase a city's population density. However, claims differ as to what methods are most appropriate, and the appropriate timespan and level of density.

The most extreme compact city policy is one of "withdrawal" from inefficient low-density settlements and towns. In this case, the government forcibly resettles all the inhabitants of those areas and terminates various services. There were, in fact, occasionally such cases where residents of sparsely populated areas were relocated *en masse* to large cities and other regions during the high-growth period.[1] The reasons behind these moves were not due to depopulation and aging people, but because residents were seeking better economic opportunities in terms of income and education elsewhere. In contemporary Japan, rapid population decline has gained ground in the regions, leading to poor efficiency in administrative services and putting pressure on local finances. Under such circumstances, withdrawing from low-density population areas has a certain rationality.

In fact, current discussions on whether to give up on such communities and proceed with withdrawal occur in earnest in districts suffering an excessive drop in population (Hayashi and Saito 2010). While talk of such "withdrawal" is still at the village or district level, the trend to nearly give up on autonomy and pursue "concentration," or rely on other larger

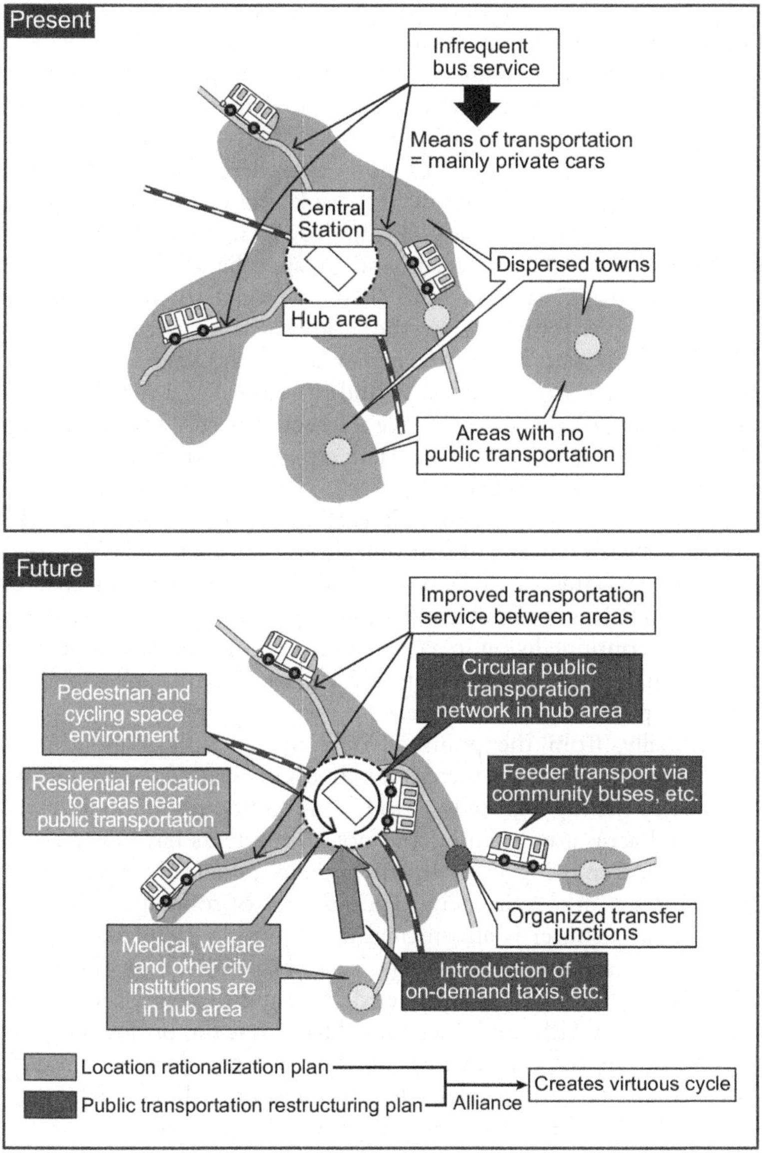

Fig. 9.1 MLIT's conceptual models for compact city. Source: MLIT, Meeting material of location rationalization planning, June 2015

nearby cities for many city services, can already be seen in many local governments. As mentioned in Chapter 3, the number of services a local government can support is limited by the size of its population. When the population shrinks, public services, such as hospitals and high schools, are difficult to sustain financially. Private services, such as commercial facilities and gas stations, will be forced to close, and there is a greater possibility that public institutions will be forced to look after residents' shopping and refueling needs. Of course, not all small towns and villages are suffering population decline. Many towns and villages have been successful in attracting "U-turns" (people who wish to return to their hometown), "J-turns" (people returning to the hometown of their spouse), and "I-turns" (people who move to a rural area but who have no original connection). In part, these "returnees" find the appeal of a rich natural environment and country living attractive. However, hardly any of these towns and villages are able to provide a certain city-scale and the equivalent city-level services in their own area.

Faced with inexorable decline, can each local government proceed with the ultimate "withdrawal and concentration" of forcibly resettling people from sparsely populated areas and conglomerating them in an area of greater density? It is impossible in today's Japan to forcibly relocate people who wish to continue living in a certain locale. And, as long as people live there, no matter how inefficient, local governments are supposed to endeavor to provide public services.

Additionally, from the point of view of residents, conditions have become such that they are forced to rely on government administration. In contrast to the high-growth period, community ties in modern villages even in rural areas are declining and not all residents have lived there for many years. After living in large cities, many people do return to their ancestral homes after retirement. The number of residents with no local connection whatsoever is also increasing alongside "I-turns." When the ties between residents diminish, the services provided by the residents themselves through mutual assistance decline and therefore the need arises to augment them with public services. Moreover, the progress of motorization, where almost everyone has his/her own car, also makes it difficult to keep residents concentrated. Roads have been built deep into the mountains so that the number of villages with limited access to cities keeps declining. Few are confined to their home region due to limited access, as was the case in the deep-snow villages centuries ago. A marginal location is no longer a valid reason for suggesting that residents relocate.

The compact city policy actually implemented took a much more resident-friendly approach and was spread out over a long time. Let us look at Toyama City in Toyama Prefecture, located in Western Japan and facing China. It is well known as a compact city. Toyama is currently engaged in dozens of measures for concentrating cities. However, city authorities are, of course, neither forcibly relocating residents from outlying areas to the city center nor particularly regulating new transfers into peripheral areas more strictly than other local governments. Its measures are centered on the inductive approach of making living in the city center being beneficial to residents. For example, in terms of infrastructure development, the light rail lines and the central commercial area it has developed are drawing national attention. In addition, the city is pursuing a chain-type structure to link multiple centers, providing a subsidy to people moving close to stations and bus stops to reduce overconcentration in single areas (Awashima 2009). The priorities considered by Toyama City over the past few years for the development of such capital investment as road and bridge development and the consolidation of public facilities as well as snow removal, may lead to a policy restraining people from living in sparsely populated areas or the city's outskirts. Although these policies presently affect the residents of limited areas, it does not mean that public services will be stopped. The compact city policy of Toyama City is used alongside resident-friendly methods for implementation so that residents' rights are respected.

If aggressively applied, would any of these moderate measures have immediate impact in any local government? Given that they are strictly inductive measures, there would be people who, turning a blind eye because of some inconvenience, would like to remain where they are. Many would be the elderly that wish to spend their remaining days in a familiar place. In that case, districts dealing with extreme depopulation would also have to simultaneously enact measures for their elderly residents, and as a result resources for government services in those districts would be insufficient. Even if measures were successful in changing the mindset of urban residents and if many chose to live in the city center, it would take decades to develop the general images people have of a "compact city." This is because in many cities throughout Japan, residential developments are still being constructed in low population density areas and government urban planning officials seem unable to stop it. Consequently, the population at the outskirts hardly falls at all, and government and the private sector will be forced to spend another generation streamlining city services. In the end, even if a moderate compact city policy was promoted, there is a strong probability that the various problems accompanying this aging and population decline will sprout up everywhere before such a vision can be realized.

However, standing back and doing nothing will definitely make the problem worse as inefficiency in the cities will increase. In order to avoid such a situation, Japan needs to steadfastly promote the following points using as a basis policy that is friendly to residents.

1. Aim at an efficient, concentrated urban structure over the next 100 years with a clear plan toward the future.
2. Address the birth rate so that a recovery that will apply the brakes on the declining population occurs in each city. This plan also has to be conceived to last for a period of 100 years.
3. Set an appropriate area so that extreme population movements do not occur, and prepare key industries/employment and urban services in that area.
4. Conduct transitional policies until a concentrated urban structure and a recovery in the birth rate have been achieved.

To achieve these goals by the year 2100, when children born now will be elderly and over 75 years of age, a variety of individual transitional measures during the twenty-first century, which are premised on poor urban structure efficiency, will need to be implemented. Regarding access to city services, for example, the basic direction should be to create an environment by 2100 where all people can access city services either on foot or by public transportation such as trains, Light Rail Transit, buses, and new types of vehicles. However, while the urban sprawl of the twenty-first century remains, residents left behind on the outskirts should be provided with transitional services such as shared taxis even if the cost is high.[2] Concerning these residents, who are at the heart of the urban problem, the residential rights of existing dwellers should be acknowledged and support for living will continue to be provided, but no new residences should be allowed. Alternatively, districts could be divided into those with full services and full residential taxes and those without in order to reduce the delivery costs of government services. In doing so, population in areas with inefficient city services would gradually dwindle and reach zero in the long term.

If Japan's unprecedented aging can be overcome by applying measures such as those mentioned above, stable, low-cost city services could be realized in the twenty-second century even in regional cities with higher population density in compact cities.

9.2 Need for "Wide-Area Cooperation" and "Local Political Reform"

Maintaining regional living areas with a thorough overhaul of the local government system and finances as well as a strong push for wide-area cooperation

To achieve this goal the following must be executed without hesitation.

We need to change how local allocation tax grants to the local governments are calculated to increase incentives for regional cooperation.

Local governments are where representatives of residents carry out decisions that play a central role in developing regional cities to cope with population decline and aging. However, some reform is required.

The local financial system, especially local allocation grant needs reform. Local governments should carry out operations with their own tax revenue and raise the efficiency of their investments. By reducing operations performed by local authorities, it will be possible for the central government to reduce its financial guarantees. However, as described in Chapter 6, reforms underway since the 1990s have essentially failed to produce the initially envisioned effect; and to the contrary, have reduced the strength of some local governments.

Local allocation tax grants to the local governments were originally determined according to the "amount of basic fiscal demand" for standard administrative services of local governments. Local governments have their own financial revenue to provide the services, however, when the revenue would not meet the fiscal demands, local allocation tax grant supplies the deficiencies.[3] In other words, the financial resources for -standard administrative services- are guaranteed by this financial system. Therefore, reducing local allocation tax grant leads to a reduction in the basic financial needs and standard services of a local government. As a result of the reforms advanced in that direction under the Koizumi Administration from 2001 to 2004, the local allocation tax grant was certainly reduced, but at the same time they also reduced administrative services in local authorities with shaky finances. The calculation mechanism needs to be altered to avoid similar situations in the future and to lessen the backlash from local governments.

One resolution might be amending and providing more generous grants to local governments that operate city services efficiently in cooperation with other local governments. Specifically, of the various coefficients used for calculation, those that are relatively advantageous to local government with low populations and population density need to be substantially overhauled. On that basis, advantageous coefficients should be freshly incorporated for local governments that work hard to streamline operations and carry out various measures such as providing city services or regional economic activity, for example, through wide-area cooperation. This would likely act as an incentive for the formation of local living areas consisting of several municipalities. "Wide-area cooperation" here does not mean multiple municipalities "merging" to become one. Each municipality would share city services with no change in its respective administrative boundaries. This method would also therefore be a counterproposal to residents who oppose mergers. In addition, such a move would also prevent other harmful effects of municipal mergers, such as the loss of regional identity or a drop in convenience and the decline of outlying districts.

Fiscal measures via the local allocation taxes can already be seen in measures for "wide-area cooperation" in existing policy, but the impact is only negligible at the moment. The Regional Central City Concept launched by the government in 2014 also seems unlikely to significantly improve the efficiency of city services, judging at least by the content of model projects to date.[4] When aiming for greater changes the key will be fiscal adjustment between those municipalities that cannot administer "standard administrative services" without local allocation tax grants and economically strong municipalities with "surplus resources" such as Tokyo that can operate services without the grants.

It is obviously inequitable to simply have the two groups compete. To ensure fair competition between local governments, the central government should stipulate what comprises "standard administrative services" and guarantee the requisite funding, and then use fiscal adjustment between local governments for financial resources to be used for purposes other than these. Without such adjustment, local governments with stronger economic powers will obtain an advantageous position, and economically weak local governments will continue to rely on the central government.

Amend laws and systems regarding local governance—municipalities and local living areas should flexibly set regulations for administrative and political/electoral systems

Under a nearly uniform electoral system and establishing a relationship between local executives and assemblies nationwide, various local authorities are criticized as they protect vested interests of those that remain in power, averse their eyes from the challenge of population decline, and refuse to implement new ideas.

A drastic overhaul of the political and electoral systems is necessary to put creative ideas that will impede Japan's population decline into action, to maintain and revive regional living areas. The most important thing is to change the electoral system for local assemblies into one where lawmakers pay attention to the wishes of many residents and directly confront the population shrinking. When local governments make important decisions, they cannot ignore assembly members. However, there is a risk that local assemblies will turn into scenes for bid rigging between members who only have narrow support bases. Since they are elected under the single non-transferrable system, where the number of votes required for winning is small, it is easy for assembly members to strengthen relations with particular groups of voters. For example, in a local assembly election with about 20 seats, winning only requires less than 5% of the vote, and if the turnout is low, even the support of 1% is enough. Naturally, candidates strengthen ties with organizations that will support them (labor unions, residents' association, alumni, etc.) and work as their delegations. As a result, some local assemblies are regarded as unnecessary, raising doubts about their independence.

To change this situation, we need to divide electoral districts even at the level of municipal assemblies, or introduce proportional representation electoral system so that specific interest groups do not form the majority. If there are many candidates per electoral district, not only the threshold for winning is low, but voters are also overwhelmed by choices and thus become unsure how to cast their votes. In addition, people who share certain interests form blocs. If representatives are sensitive to the voices of many voters, and if seat numbers are managed appropriately, it is likely to encourage investment for updating into an efficient city.

One of the causes for insufficient investment is in urban centers and the continued unregulated expansion of city suburbs, or city sprawl, because local authorities tend to be conscious of the interests of people who are rooted in the region rather than paying attention to those who relocate in search of amenities.

Another important aspect for regional political reform is creating a political force (a political party) that will be responsible for decision-making over an area wider than the municipal level. Given the estimates for future population decline, even if each municipality concentrates investment in the center to redevelop the town, investment as a whole would become excessively dispersive if all local governments did the same. This raises the need to pool resources by encouraging area residents to live in a wider area beyond the framework of a single municipality. However, when local heads and assembly members hardly ignore specific groups, they may not take the decision that surpasses the boundaries of local government. People need to be organized who not only care about their own local district, but also emphasize building a highly convenient regional network where they would like to settle. For that purpose, a centralized political party that can take difficult decisions from a broader perspective, such as a withdrawal from depopulated areas in some cases, while keeping an eye on the interests of small communities is needed. Only when such an organization is formed are cities more likely to be developed into attractive places.

The "full-set-principle" way of thinking, which means local government provides everything by itself, and has a sense of belonging to the municipality, is very strong among local government lawmakers. However, many routine public services such as the management and operation of public facilities relating to "building things," are more efficient economies of scale when integrated for a large population. It has become increasingly difficult for small regional municipalities especially with significant population decline to maintain a variety of facilities and services alone, and joint management within a local living area will be beneficial.

In practice, however, multifaceted cooperation between local governments has made little progress in Japan. There are hardly any local governments currently contemplating cooperation with neighboring municipalities on public facilities, with the exception of facilities that are necessary but undesirable in the vicinity of residential areas such as water/sewage and garbage incineration, industrial waste treatment facilities, crematories, and sewage treatment facilities. At best, only shared library use is being promoted. On the other hand, residents, government officials, and private operators living in regional areas have a strong sense of crisis about the issues of depopulation. Despite a population decline being covered by state and research institutions since the early 2000s, most municipalities assumed their populations would maintain equilibrium or increase (Taira 2005). By the late 2000s, however, the concrete problems of population decline described in this book became increasingly clear and dramatically changed local government perceptions of the future.

A typical initiative that illustrates this problem is the integration of public facilities. The large number of schools, libraries, cultural halls, sports facilities, community centers, and other institutions that were built prior to the 1970s during Japan's period of high population growth, are starting to outlive their usefulness as the phase of population decline approaches. Unless all of these functions are integrated, the costs of maintaining, upgrading, and earthquake reinforcing will mount, forcing local government into inefficient use of their budgets.

From the late 2000s, innovative local governments have been tackling problems such as these based on public facilities white papers, which analyze the situations of all public facilities. In some places, like Hamamatsu in Shizuoka Prefecture, where the duplication in public facilities was obvious to everyone after a large-scale merger, local governments are rushing to integrate and raise service efficiency. Local governments are improving the efficiency of public facilities in preparation for future aging and population decline, such as in Hadano in Kanagawa Prefecture, where the population is not currently falling. Some of the consolidated facilities have been transferred or lent to private businesses and local residents and are being used effectively.

Still, there are many local governments that cannot make headway with consolidation due to strong opposition from the delegations of specific interest groups in local assemblies or a fearful negative attitude from the heads of local governments. Other background factors are numerous institutional barriers to consolidation and weak financial backup. Consequently, even if a local government manages to publish a public facilities white paper, many cannot subsequently come up with an integration policy. In situations such as these where response by a single local authority is difficult, it is even more politically difficult for a regional living area comprising several municipalities to deal with them. Creating a centralized political party in a broader alignment is crucial when dealing with these problems.

Reviewing the uniform local government political institution should also be considered. Directly elected head and a large number of local assembly members are not necessarily required in each government. In Europe, where the history of local government is much longer than in Japan, systems not only differ between countries, but also differ significantly by state, county, and municipality even within a single country like Germany. Political institutions in Japan could also differ according to area (metropolitan or rural) and city-scale, geographical conditions, or to such differences as major industry, history, and culture. For example, in local governments where population decline is severe and "selection and concentration" are required, town-planning professionals could be invited in as a town manager in charge of revitalization strategy. Some local governments with severe municipal finances have reduced salary for the assembly members without reducing their number; it is also possible to assemble volunteer members, and instead welcome many more of them to contribute a variety of opinions and ideas. In contrast, areas where sophisticated measures and urban development are required from an international perspective, such as central Tokyo, could be managed directly through central government control.

In Western countries, a variety of different political institutions are well established throughout nations such as the USA and Germany. Japan seems to lag behind Western countries in its reforming government system. However, when each region is required to respond to the situation of population decline, it is necessary to change Japan's local government system as well.

9.3 Considering the Minimum Line for Population Density

How can Japan control and direct population, households, and urban density? Regardless of whether the population is falling or rising, this is the fundamental challenge in guiding cities and living areas toward a better direction.

Urban planning has tackled this issue primarily in two ways: "land use control" and "urban facilities development." Having ensured convenience by placing various urban facilities in the city center, the fundamental principle in forming and engendering high-density urban areas is to create a network linking those facilities to the suburbs with linear social infrastructure (transportation such as roads, railways and bus, water and sewerage, electricity and gas). Weighing their mutual relationship, the supply of city services has furthermore been pursued by developing and maintaining public facilities, transportation and water, as well as land use control and guidance including density regulations on building heights, floor space, and zoning.

However, this relationship was not successfully controlled in Japan. A particular problem is land use control, but especially population density control. Entering a depopulation phase has made this problem increasingly serious.

When the population falls, so does the total amount of human activity. Of course, the required volume of land use also falls, but land use planning and regulation have not kept pace with this trend at all. Regulations remain lax and the maximum volume of land and floor space that the city as a whole can use is still set excessively high. While deregulation and economic activity should probably be encouraged in a pinpoint fashion for a very small number of districts where demand is expected to be high even in a phase of population decline, it is undesirable for regulation to be lenient in a given area as a whole. As a result, cities have sprawled, and population density is low and continues to fall.

The minimum population density for efficiently supplying city services is said to be around 40 persons per hectare. It is also one of the criteria used by local governments when allowing development plans to urban areas. The results of a survey in Toyama show that 40 people per hectare is the dividing line for population density; above this residents shoulder a greater burden than their benefits, and below this residents enjoy greater benefits than their burden (Fig. 9.2).

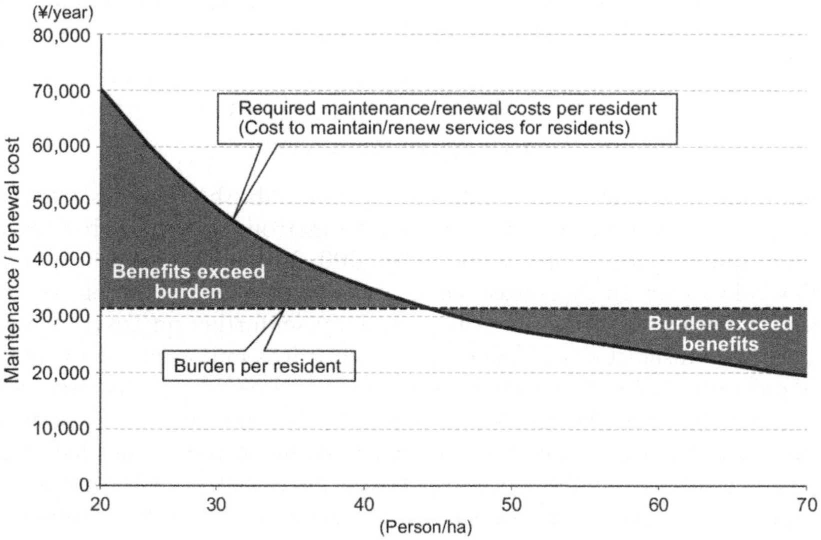

Fig. 9.2 Relationship between population density and administrative cost per resident (maintenance/renewal) in Toyama City. Source: Compact City Study Group, 2004

It is possible that city services such as buses and other forms of transportation, water and sewerage supply, garbage disposal, nursing care, and childcare are all inefficient in cities and prefectures that are either level with or below this threshold because residents are relatively diffused even downtown. If in the future the population in these districts falls even further, inefficiencies will increase and a fear of being unable to maintain services will emerge. Viewed from a national perspective, the level of population density is at a level that cannot be overlooked. In Japan, originally a mountainous land with many uninhabitable areas, urban areas where many reside have been defined as "demographically intense districts" (DID). These are "in principle, an area of 5000 people or more with adjacent districts with a population density of 4000 people or more per square kilometer (which equals 40 people/hectare)." The urban level matches exactly the population density mentioned above, but as the population falls, it will barely meet this threshold. Districts where the level deviates from the DID will no longer be classified as urban.

Looking at Fig. 9.3, there are many places, even entire prefectures, which are already approaching the level of 40 people per hectare. In the near future when depopulation takes off in earnest, it is expected that many districts will no longer be urban areas.

9.4 Medium-Sized Cities as "Population Reservoirs"

Medium-sized cities around major population hubs should act as a "reservoir" preventing an outflow of people to metropolitan areas from rural areas, but even local cities of about 100,000–200,000 people fail to fulfill this role today. In Northeast Japan, for example, young people bypass local cities heading to Sendai with their sights set further on Tokyo. Under these circumstances, it is argued that even if it is impossible to maintain population in all municipalities, maintaining and reviving prefectural capitals and cities one size smaller—where there is some prospect of population maintenance—should be supported. While opinion is divided as to whether development in places like Tokyo and Sendai could be artificially suppressed or not, there is a shared awareness that a limited number of metropolitan areas are swallowing up people as Japan's total population falls. Regional urban areas supporting them can act as reservoirs and would also be backed by people living in regional areas.

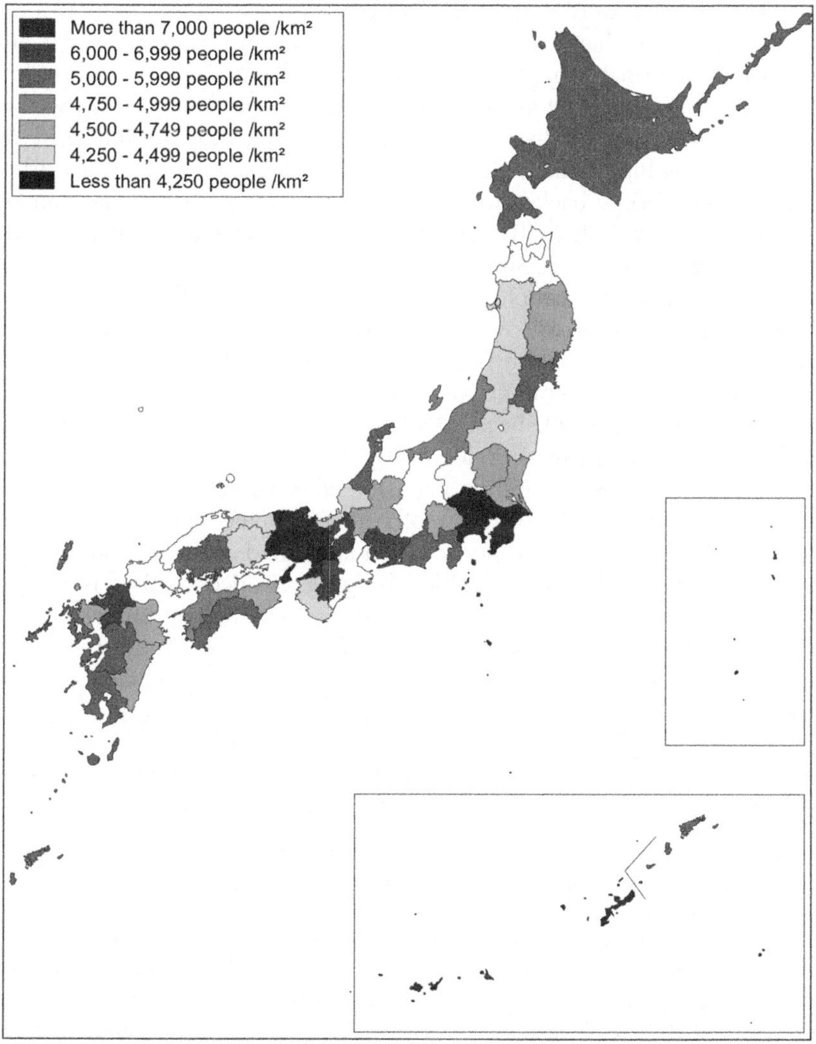

Fig. 9.3 Population intensity of demographically intense districts (DID). Source: MIC, Population and Households of Japan, 2010

The continual outflow of people to prefectural capitals and metropolitan areas might change if, with assistance, education, employment and welfare in regional urban areas could be bolstered. This is because it is expected that even if people cannot remain in their isolated villages, they could through support live in towns relatively near to the countryside. In addition, it is highly likely that the support of promising regional urban areas would also be backed in the Tokyo metropolitan area. The number of elderly will explode in Tokyo in the future, causing a shortage of medical facilities such as nursing care, welfare facilities, and hospitals. In response to this, for example, as mentioned in Chapter 2, Suginami Ward, Tokyo, and Minamiizu Town, Shizuoka Prefecture, agreed in 2014 to jointly establish a special nursing home for the elderly. Such a home must be established in each municipality by the local authorities, by law. This arrangement was possible because Suginami Ward's need for land to create new facilities for its increasing elderly population coincided with Minamiizu Town's land availability and its need for financial support, employment, and plans to prevent the outflow of young people. However, as seen in Chapter 2, whether this method can be maintained in the medium to long term even in regional municipalities, as aging advances further, is unsure.

Even if the case of Suginami is an extreme one, it is no longer unusual for the elderly in municipalities in the Tokyo area who cannot be accommodated to be sent elsewhere. If suitable sites for hospitals and welfare facilities cannot be found in dense, expensive city centers and suburbs, discussions will likely turn to utilizing regional areas. Many elderly people might relocate of their free will if moderate facilities and city services are set up in such regional areas with a rich natural environment.

9.5 Seeking a Revival in Community

Finally, this chapter makes some recommendations on crisis management with regard to overconcentration in Tokyo. With advancing demographic aging in the Tokyo metropolitan area, the number of residents in need of disaster assistance, should one occur, is increasing, but ties between residents are weakening in local communities. Furthermore, an increase in single people and foreigners also raise the hurdle higher for lifesaving in emergencies. Moreover, the central functions of politics, government administration, and the economy are very highly concentrated in Tokyo, leading to concern as to whether the capital's functions could be maintained if it were hit by a large-scale disaster. The following subsection looks at the necessary measures for the elderly and foreigners in the metropolitan area, the capital function, and social and economic functions.

9.5.1 Elders and Foreigners

The elderly and foreigners are listed in the Disaster Countermeasures Basic Act as "persons requiring assistance" in a time of disaster because it is difficult for them to single-handedly obtain the necessary information quickly and accurately as well as to evacuate to a safe place. For the elderly, who will continue to increase, it is important that each local government in the metropolitan area proceed with the following four points.

1. Improve and update name registers to gain awareness of people requiring assistance and share that information with related parties. Also decide in advance how the register is to be handled.
2. Prepare a variety of means to communicate necessary information to people requiring assistance and build a local support system connecting them.
3. Develop a support system that takes into account a variety of peoples requiring assistance such as the elderly and people with disabilities.
4. Promote designation and agreements for welfare shelters, and develop a system in cooperation with facility operators and non-governmental organizations so that they can be opened and operated quickly and smoothly in the event of a disaster.

An additional approach would be to concentrate elder facilities in areas that are resilient to disasters and crises such as areas on strong ground. Most disaster victims in Japan from earthquakes, wind and flood damage, or heavy snow are over 65 years of age. Based on this statistic, both the national and local governments have been promoting disaster countermeasures for people requiring assistance as a key issue. However, the Great East Japan Earthquake revealed the insufficient nature of those measures. While there were some elderly people who had ties with local residents on a daily basis and whose circumstances were grasped smoothly, the safety of many had to be checked door to door on the basis of government registers by social workers and neighborhood associations. Confirmation was also difficult for residents not listed in the register due to privacy considerations over personal information.

This earthquake also highlighted other problems for people requiring assistance, as some had trouble staying in evacuation shelters such as gymnasiums that were not barrier free and should have taken refuge in welfare shelters. However, the elder welfare facilities designated as a shelter were themselves damaged by the earthquake, or suffered bed, food, fuel, and staff shortages with the sharp jump in evacuees, many taking a considerable amount of time to fully function. There were also situations where facilities struggled to operate because roles were unclear for providing practical assistance to the evacuees. Moreover, with lifelines cut and a difficulty in securing fuel, it was not always possible to fully support the needs of people needing dialysis and ventilators.

There is a strong likelihood that these problems will occur even in the metropolitan area with its rising number of elderly and care recipients, and could lead to significant confusion. Many of the elderly either never use or are not good at using computers or mobile terminals to gather information. There is a concern that they will become “information refugees” in times of a disaster. Today, when the Internet is widely used as one kind of infrastructure, evacuation information, contact about goods distribution, and announcements on a variety of consultations will mainly take place online. Information via posters and the radio will probably remain but to a much lesser extent when compared to the frequency of updates and the amount of information online. Accordingly, the fear that elders living alone will end up in a situation where they cannot obtain the information required to survive is especially high.

As described above, there are many places in the metropolitan area where local communities have not been formed. It is possible to reduce the human suffering during times of disaster if local residents check each other’s safety and help each other. Therefore, it is important to increase the number of local inhabitants with a sense of autonomy. In the Great East Japan Earthquake, government institutions carried much of the life-related support. However, in some regions, town halls and other government agencies themselves were affected, and they were unable to mount an adequate disaster response. In circumstances like this, there was a difference between evacuees in areas with a sense of autonomy and those without. For example, in the former areas, residents acted voluntarily in a spirit of “mutual help,” but in the latter areas where residents relied on government agencies, there were cases where relief supplies were not distributed efficiently. For this reason, the importance of self-government organizations needs to be reaffirmed so that as many residents as possible take action to help each other in times of large-scale disasters.

Iida City in Nagano Prefecture and Toyama City are two useful examples for strengthening local capabilities. Iida City is promoting town development not only according to a government-formulated plan, but also in line with a basic concept made by residents. As a result, a sense of autonomy and community attachment is strong among local residents. Also, a local autonomous organization helps foreign residents deal with the challenges of everyday life, such as garbage separation. This has resulted in a higher degree of involvement between foreign residents and the autonomous organization, with foreign residents for example participating in disaster prevention drills. Non-Profit Organization (NPO) activities such as education and foreign resident aid also thrive in the City. Since NPOs are active in supporting disaster victims, this is also desirable in terms of disaster preparedness.[5]

On the other hand, Toyama City, as mentioned already, is working on a compact, centralized urban development based around a public transportation axis. Promoting a compact city has enabled neighborhood interaction to be maintained, thereby preventing a decline in local disaster prevention organizations associated with a drop in population density. As a result, in the event of a disaster, local residents can be expected to help each other. Making the city compact also reduces transport times for ambulances or firefighting, increasing the number of survivors.[6]

9.6 Policy Measures for the Worst Scenario

9.6.1 Capital Functions

Both the national and local governments are working individually to ensure the continuous functioning of the capital in times of large-scale disasters, but it is not sufficient. The following measures are required in terms of backing up capital functions.

1. Secure Backup Facilities
   Prepare conference facilities large enough to hold parliament, emergency disaster response headquarters, and office facilities for each ministry and agency both inside and outside the metropolitan area. At the same time, back up the data required for the country's business continuity.

2. Means of Communication with Tokyo—Ensuring the Transportation Network
   Ensure the communication and transportation between central Tokyo, where capital functions are congregated, and the backup facilities that will not be affected by disaster or that can be quickly recovered if damaged.
3. Partial Decentralization of State Functions
   Place a part of crisis management functions in normal times outside Tokyo to ensure the safety of personnel responsible for running state operations.

Legislative functions such as the National Assembly, the executive functions of the Cabinet and the central government, and the judicial function including the Supreme Court are all concentrated in Tokyo. If a large-scale earthquake hit Tokyo, these three functions would all immediately be subjected to damage, raising the fear that the nation would fall into crisis. Taking into consideration the damage to people and organizations responsible for each of these functions, information and communication, lifelines, and transportation, the impact of overconcentration is serious. In the government's National Resilience Plan, Hokkaido is currently being considered to play the emergency role of backup if it becomes impossible to maintain central functions in the metropolitan area. However, concrete discussions at the national level for the temporary relocation of the capital have not made any progress.

9.6.2 Social and Economic Functions

The social and economic challenge is the concentration of major companies' headquarters in the Tokyo metropolitan area. Each company is promoting measures such as developing business continuity plans, and the government also prompting corporate efforts by issuing guidelines. The Business Continuity Guidelines (Third Edition, August 2013) issued by the Cabinet Office (Disaster Management), cites the following as important elements in the business continuity strategy of companies and organizations[7]:

1. Supplying continuity and early recovery of important products and services
2. Securing central functions of companies and organizations
3. Maintaining information and information systems

4. Securing funding
5. Responding to laws and regulations
6. Ensuring consistency with administrative and social infrastructure business initiatives

Many companies have already taken measures such as designating agents, ensuring an alternative base for headquarters' central functions, and data/document backup. In many cases, alternate bases are in their own facilities, such as branch offices. Moves can also be seen to decentralize head office functions away from Tokyo. But transferring functions is not sufficiently advanced for companies whose alternate facilities are in high disaster risk areas. The state also needs to develop land and infrastructure to allow companies to relocate to low-risk areas.

In this way, Tokyo, like many other capitals in the world, includes both the political center and the economic center. Its outsized population in domestic terms is similar to London, Paris, and Seoul. This is in contrast to relatively small capitals that focus on political functions such as Washington, DC, and Brasilia. In addition, the concentration of economic functions in the Tokyo area during normal times brings with it the benefits of integration and diverse functions being performed efficiently. Especially in the economic field, Tokyo has established itself as an important business hub and is proud of its huge economies of scale even in global terms. However, Japan has experienced many disasters in the form of earthquakes, tsunami, and typhoons, often suffering extensive damage. A large-scale earthquake is also predicted to occur in the Tokyo metropolitan area, and there is fear that the human suffering, building damage, and economic damage would be extremely serious.

At the end of 2013, the Central Disaster Management Council released damage estimates if a magnitude 7 earthquake occurred directly beneath Tokyo. The human damage was predicted at a maximum of 23,000 dead and social/economic damage at 95 trillion yen. However, these predictions are "within assumptions." For example, there is concern that the number of casualties would be far greater during the morning and evening commuter rush at railway stations. There is also the example of the Great Kanto Earthquake of 1923 and the Tokyo air raids during World War II, where multiple fires occurring simultaneously turned into a so-called "fire whirlwind," greatly increasing the damage. Despite the presence of dense areas of wooden buildings in Tokyo, no consideration has been given to these.[8] If a typhoon or strong winds occurred at the time of an earthquake, the damage is likely to be expanded to 100,000 casualties.

Regarding economic damage, damage that cannot be quantified such as loss of data or loss of asset value due to falling land prices also have not been taken into account. Therefore, in a worst-case scenario, the damage would be much greater. The final report of the Expert Committee on Earthquake/Tsunami Countermeasures from Lessons Learnt in the Tohoku Pacific Ocean Earthquake proposes: "when making assumptions on earthquakes or tsunamis in the future, a huge earthquake of the maximum class taking into account all possibilities should be considered."[9] Future measures for the Tokyo metropolitan area need to be based on the experiences of the Great East Japan Earthquake and bear in mind that the unexpected always happens.

Notes

1. According to the Survey of Municipalities concerning the Restructuring of Rural Communities, there were 87 municipalities, mainly in the Hokuriku and Shikoku regions, with rural communities that had lost their entire population as the result of voluntary migration by their residents during or after 1965. During the same period, there were 65 municipalities, mainly in the Hokuriku and Tohoku regions, with rural municipalities that conducted planned relocation. See Chaps. 5 and 6 in Rural Development Planning Commission (1992).
2. Toyama Transportation Office n.d.
3. Basic financial revenues do not correspond to the full amount of tax revenue that is assumed for each municipality, but is set at 75% thereof. The remaining 25% is to be used as reserve financial resources for specific municipal demands so to speak in excess of standard financial demands.
4. The Regional Central City Concept is one type of wide area coordination between municipalities that is conducted within the urban scope of central cities that meets national requirements. An urban region larger than a settlement and independence area in terms of population and area is envisioned. Not only daily services but also industrial development, among other things, are included in the potential scope of the coordination according to report from the Ministry of Internal Affairs and Communications (MIC 2014).
5. Mayor Makino Mitsuo and Multicultural Chief Ujihara Rieko, personal communication, August 12, 2014.
6. Mayor Mori Masashi, personal communication, October 29, 2014.
7. Cabinet Office 2013.
8. Disaster prevention working group 2013.
9. Cabinet Office 2011.

References

Awashima, Yasuo. (2009). The reasons Toyama City aimed to build a Compact City (Toyama shi wa naze konpakuto shiti wo mezashitanoka). Presented at the 3rd Compact City Promotion Study Seminar. http://www.thr.mlit.go.jp/compact-city/contents/suishinkenkyuukai/3/s1.pdf, accessed on September 23, 2016 (in Japanese).

Cabinet Office, Government of Japan. (2011). Report of the committee for technical investigation on countermeasures for earthquakes and tsunamis based on the lessons learned from the "2011 off the Pacific coast of Tohoku Earthquake". http://www.bousai.go.jp/kaigirep/chousakai/tohokukyokun/pdf/Report.pdf, accessed on September 23, 2016.

Cabinet Office, Government of Japan. (2013). The business continuity guidelines (Jigyo gaiyo gaidorain). http://www.bousai.go.jp/kyoiku/kigyou/keizoku/pdf/guideline03.pdf, accessed on September 23, 2016 (in Japanese).

Disaster prevention countermeasure for earthquake in the capital working group. (2013). Central disaster management council documents (Chuo bosai kaigi). http://www.bousai.go.jp/kaigirep/kentokai/hisaishashien2/pdf/dai5kai/siryo3.pdf, accessed on September 23, 2016 (in Japanese).

Hayashi, Naoki & Susumu Saito, eds. (2010). *Rural planning of withdrawal* (*Tettai no noson keikaku*). Tokyo: Gakugei Shuppan (in Japanese).

MIC. (2014). Concept and promotion guidelines of establishment of regional central city concept in metropolitan area (Chihoshuchutoshiken koso suishin yoko no gaiyo). http://www.soumu.go.jp/main_content/000309074.pdf, accessed on September 23, 2016 (in Japanese).

MIC, Statistics Bureau. (2010). Population and Households of Japan (Summary report and statistical tables, 2010 population census). http://www.stat.go.jp/english/data/kokusei/2010/summary.htm, accessed on February 27, 2017.

MLIT. (2015, June 1). Location adequacy plan briefing session document (ricchi tekiseika keikauk setsumeikai shiryou). http://www.mlit.go.jp/common/001091253.pdf, accessed on February 27, 2017 (in Japanese).

Rural Development Planning Commission. (1992, March). Extinction of Settlements and Ruination of Agricultural Land in Depopulated "ChuSankan" Areas. *Rural Engineering Study* (*Noson kogaku kenkyu*) 54, 45–82. http://www.rdpc.or.jp/contents/01kenkyu/pdf_kenkyu01/54.pdf, accessed on September 23, 2016 (in Japanese).

Taira, Nobuhisa. (2005). *Falling population measures required of the regions* (*Chiiki ni motomerareru jinko gensho taisaku*). Saitama: Seigakuin University Press (in Japanese).

Toyama Transportation Office. (n.d.). Community bus and shared taxi (Komyuniti basu, noriai takusi). http://wwwtb.mlit.go.jp/hokushin/toyama/bus_taxi/com_bus1.html, accessed on September 23, 2016 (in Japanese).

CHAPTER 10

Countering Falling Regional Population with Business

Matsuda Kiyoto, Arai Junji, and Nagao Takashi

A major factor behind Japan's declining population is that young people head to big cities because locally they are unable to find adequate work. For example, the number of new residents in the Tokyo metropolitan area outstrips departures by approximately 100,000 each year. The most effective way to stem this tide is to create rewarding, appropriately compensated, stable jobs in the regions. As one of the first countries to be affected by this issue, Japan should see this as a chance to show the world policy responses and to use a repertoire of successful initiatives as a guide for boldly improving the attractiveness of the outlying regions. Strong companies that are supported by the local community and prosper are not only good at making a profit, but also have mechanisms for returning revenue to their communities. In dealing with the falling birth rate, cooperation in the community and workplace will be essential when parents have babies, as well as for fostering understanding from and participation by the

Matsuda. K
Unison Capital, Inc., Tokyo, Japan

Arai. J (✉)
Product Strategies Department, Lawson, Inc., Tokyo, Japan

Nagao. T
Division of Business Strategy, Yamato Transport Co., Ltd., Tokyo, Japan

Y. Funabashi (ed.), *Japan's Population Implosion*,
https://doi.org/10.1007/978-981-10-4983-5_10

husband for child rearing. Only in such families, communities, and workplaces will young women have the eagerness for work. Currently there are more and better opportunities for women to run businesses, such as retail and services using the Internet. And robust businesses generating quality employment, energizing the local economy, and minimizing residents' anxiety about the economic future make a significant contribution to the enhancing the willingness to have and rear children. According to the Small and Medium Enterprises Agency, there is an increasing number of young people leaving the city and choosing to live in regional areas in search of a more relaxed lifestyle.[1] It is important to create employment opportunities to attract such young people from within and outside of the region to settle down in that area. Below, we examine a number of examples from across the archipelago where talented people have been enticed by new opportunities, which in turn helped activate the regional economy.

10.1 Komatsu: Aiming to Balance Work and Childcare

In May 2011, on the 90th anniversary of its founding, Tokyo-based construction machinery giant Komatsu Ltd. transferred some of the functions of its headquarters to its birthplace in Komatsu City, Ishikawa Prefecture and built a comprehensive training facility named *Komatsu no Mori*, Komatsu Village. On a vast factory site, the corporation set up training facilities and a green zone, a facility where construction machinery could be studied and actually test-driven, and the world's largest dump truck display. All of these facilities, with the exception of the training facility, are open to the general public. Komatsu's reason for erecting such facilities in its hometown was not mere nostalgia nor to appeal to local residents, but rather it was based on its determination to become involved in making the outlying region of its origin more attractive to future employees and on a careful calculation that this would ultimately lead to increased profits.

In the three prefectures that make up the Hokuriku Region (Toyama, Fukui, and Ishikawa), the number of people working in agriculture fell approximately 14% between 2011 and 2013. A decline in agriculture strips the region of its vitality and has led to an exodus of people and thus a drop in population. Under such circumstances, Komatsu signed a Comprehensive Partnership Agreement on Agriculture with Ishikawa Prefecture in 2013, which involved advice about production and business management, and how to apply it to agriculture. The following year, in conjunction with local financial institutions and Ishikawa Prefecture, they also established a fund for developing technology to improve efficiency, productivity and profitability in agriculture. Komatsu not only provided the funding, but is also involved in evaluating and developing new technology. Ishikawa Prefecture has high hopes for Komatsu's efforts because "injecting know-how from other industries into agriculture is a first in the country and is important for making agriculture more competitive and its development as an industry."[2] Sakane Masahiro, advisor to Komatsu, notes that supporting agriculture will see more workers stay in the area and we expect this will in turn help us secure a labor force for factories. Since living costs are lower in the regions, raising employment there will make it easier to maintain the competitive edge, a secondary aim.

As has already been seen in Chapter 5, even childcare support and work-life balance measures that were supposedly key demographic policies in recent years have hit numerous stumbling blocks due to, among other reasons, Japan's silo mentality in the central government. In that respect, these efforts of the private company Komatsu are potentially a way out of the trap. Looking at one example to achieve work-life balance, the company has extended childcare leave to up to three years and also installed a childcare facility in one of its factories (Komatsu Kids Oyama at the Oyama Plant in Tochigi Prefecture). The number of employees taking childcare leave rose from 74 to 85 in 2013 and included seven male workers. The company's initiatives also provide opportunities to more widely take female work styles and careers into consideration. Komatsu runs training for all female employees in their third year and has also started career plan training for those hired locally, where career follow-up was sometimes lax. Senior employees present their experiences, and explain the company's in-house system regarding promotion of a work-life balance. The venue is, of course, *Komatsu Village*.

10.2 Shonai Bank: A Producer for Regional Development

Tsuruoka City in Yamagata Prefecture was the feudal headquarters of the Shonai clan and it stands out historically because the Sakai family, the descendants of the feudal lord living in the castle town from the Edo Period (1603–1867), still lives there. However, Tsuruoka's population, just over 130,000 today, is predicted to fall to approximately 70% of its current level to 94,000 by 2040. This decline is 20 years faster than the national average. Despite the harsh prediction of population decline, private firms are taking a unique approach to create new business models in the region and the Shonai Bank has taken the lead.

An industry/government/academic consortium comprising Yamagata Prefecture, Yamagata University, Yamagata Agricultural Corporation Association, and Yamagata Agriculture Support Center was established in 2009 with the Bank's think tank, Shogin Research Institute (now the FIDEA Research Institute Corporation), serving as the secretariat. It opened the Yamagata "Sixtiary" Industry Business School for technicians to carry out an entire range of industrial processes so that companies operate production (primary industry), processing (secondary industry), and distribution and sales (tertiary industry).[3] It is also involved in the revitalization of regional cities proposed in Chapter 9, in the form of specific activities based around the Tsuruoka City Planning Corporation bringing together local businesses. In addition, spurred on by the Knowledge Network Project put forward by the Shonai Value Development Study Group—which grew out of an in-house project and comprising experts from inside and outside the bank as well as local interested parties—in 2001, Keio University established its Institute for Advanced Biosciences in Tsuruoka City. Biotech-related venture firms have emerged from here, and a fourth company is about to be launched. Business has been brisk for Human Metabolome Technologies Inc., now listed on the stock exchange, and Spiber Inc., which established a joint venture with automotive parts company Kojima Industries Corporation. Tsuruoka is gradually becoming both a center for biotechnology research and an entrepreneurial base for biotech ventures. "The idea is to make Tsuruoka into a Bio-Valley. Collaboration has already started between local companies and bioventures. It's the role of regional financial institutions to fuse the two together," noted Shonai Bank President, Kunii Hideo.[4]

10.3 Hirata Farm: Public-Spirited Support of Infrastructure

Hirata Farm Co., Ltd., headquartered in Yamagata Prefecture, is a leading manufacturer of pork in Japan and consists entirely of in-house production, processing, distribution, and sales of its two famous brands of pork.

Since its inception in 1964, it has pursued a variety of initiatives to provide tasty, safe, secure and high-quality products. The business, which initially began with two pigs, has now expanded into a group with sales on a scale of 20 billion yen annually. It is currently looking into expanding overseas. Aiming to carry on the tradition of offering safe and delicious food by raising national food self-sufficiency, second-generation president Nitta Kashichi has moved forward with the "sixtiary industry" (conducting all primary, secondary and tertiary processes in-house). He has run the business with a sense of mission to "increase the population and make the area the best; to have the highest income level in the prefecture; and finally to make the area one young people want to stay in." The idea of the founding Chairman Nitta Kaichi, "Revenue will be generously returned to the community," has been passed down even today.5 Chairman Nitta considered it essential to create "airports, schools, museums, and freeways" to avoid a population outflow and to stimulate the local economy. As a private firm, the company played a leading role in opening the Shonai Airport Building Co., Ltd., in 1991, and in founding the first four-year college in Sakata, the Tohoku University of Community Service and Science in 2001. The company has also preserved local culture and historical heritage by investing in the Sakata City Museum of Art, which annually attracts some 130,000 tourists annually, more than the city's entire population. Its restaurant Soma-Ro has been in business since the Edo Period. Hirata Farm shows that a single private company can have a major influence on its region.

10.4 Kitakyushu Yamorisha: Private Sector-Led Town Revitalization

Main commercial streets full of shuttered shops are now a common sight in towns across Japan's rural countryside. The vacancy rate in the central part of Kokura, Kitakyushu City, was 18.1% as of 2008. This is significantly higher than the national average (13.1%), and higher even than Kitakyushu as a whole (15.3%). It is a typical example of the vacant house problem covered in Chapter 2. Shimizu Yoshitsugu, who had been involved in town regeneration in Tokyo, joined the project in Kokura. Sharing the philosophy of "a modern home guardian" advocated by Shimizu, the city has implemented regeneration and urban development plans led by the private sector along with real estate owner Kakehashi Terumoto.[6] At the implementation stage in April 2012, Kakehashi and architect Shimada Yohei, along with Toya Hiroki, an incubation manager who runs a cafe in North Kokura Station, Associate Professor Kataoka Hiroyuki of the University of Kitakyushu, and Associate Professor Tokuda Mitsuhiro of the Kyushu Institute of Technology Graduate School, were involved in launching the Kitakyushu Yamorisha Co., Ltd.[7] This company operates a "renovation school" twice a year in conjunction with Kitakyushu City. Approximately 60 places are offered, and students study practical urban development during a four-day course using actual buildings as their teaching materials. In addition to raising awareness about the uses of idle real estate, the course has trained people with the necessary knowhow for regenerating an urban landscape. A number of buildings in the center of Kokura have already been refurbished through these courses. Completely renovated buildings and stores have attracted shoppers and tourists to the city. In 2013, average daily traffic in central Kokura had increased by 3000 compared to 2009. The opening of new shops and offices also created new employment. These changes have been responsible for more than 300 new jobs in central Kokura.

10.5 Nakamura Brace Co., Ltd.: An Innovation Firm That Attracts Young People Even to the Countryside

Prosthetics and orthotics manufacturer Nakamura Brace is a typical small- to medium-sized enterprise with just under 70 employees. However, it does differ in one regard from your average small business: Nakamura Brace is located in Ohmori Town, Ohta City, Shimane Prefecture, which only had 403 inhabitants as of April 1, 2015. A company that earns 1 billion yen annually is unusual to be situated in such a tiny village. Underpinning the development of this unique company is its advanced technology. It manufactures artificial breasts, limbs, and other body parts from silicon. It calls its products "medical art" and trains in-house the "medical artists" who craft these products. Orders pour in not only from Japan but also from overseas. The company, which President Nakamura Toshiro started on his own in a barn in his hometown in 1974, has grown into a global enterprise. Nakamura Brace's workers are not just locals but it is staffed with young people from all over Japan. His reliable technology ensures profits, enabling both generous worker welfare and product development. The social significance of the business itself also seems to be another reason why young people want to work there.

10.6 Saku Central Hospital: Local Preventive Medicine

From the moment of his appointment in 1945, former director of Saku Central Hospital (Saku City, Nagano Prefecture), the late Wakatsuki Toshikazu was engaged in medical activities closely linked to the lives of the local people. The best example of this would be the medical check-ups available for all villagers, designed to assist in local "preventive health care" in cooperation with the local government (Yachiho Village, now merged into Sakuho Town). Wakatsuki's personal philosophy about supporting a healthy local community is still alive at the hospital today. For example, every day a "check-up corps" is dispatched from the group facility throughout the prefecture and the Japan Agricultural Cooperative (JA) Nagano Koseiren Health Management Center conducts group health screenings.

As mentioned in Chapter 2 and 3, there may be a mismatch between the distribution of the elderly population and medical and nursing care services in rural and urban areas. In such circumstances, it is not only Nagano Prefecture that needs to keep alive the idea of "preventive health care." Preventive health care is linked to primary care, which is currently attracting national attention. Primary care refers to medical services centering on local community doctors who deal with day-to-day health problems. In Japan today, 90% of medical expenses are used for highly specialized medical care (secondary and tertiary care), with only some 10% allocated to primary medical care. However, bearing in mind rising medical expenses and the shortage of regional doctors, primary care will need to be more closely monitored. With its medical check-ups for all villagers, Saku Central Hospital could be said to have practiced primary care in an era when the term was unknown. Doctors responsible for primary care are referred to as "general practitioners," and the Ministry of Health, Labour and Welfare (MHLW) is gradually aiming to increase their numbers. A 2013 report by the ministerial Committee on Specialists announced that it was working on the certification and training of "general medical specialists," stating that it is also important to have 'community doctors' able to respond to different needs depending on the region.[8]

Currently, Nagano Prefecture has its own training for general practitioners involved in area health care activities. The training program at Saku Central Hospital's Regional Medical Department has been certified as one of these general practitioner training programs. Adherence to and development of Wakatsuki's philosophy about local health care is also being promoted as part of a town planning initiative focused on the hospital called "Medicopolis."[9] This new facility aims to break away from merely emphasizing the specific needs of agricultural areas such as improving farmers' health by investigating local epidemics and offering treatments on the regional level with a focus on health care. It is essential to foster industry by enhancing welfare and medical systems, and to improve educational facilities for healthcare human resources and medical care. These changes are taking shape in the form of bringing a medical and nursing university to the area and opening a clinic (Komi Clinic) in the conveniently located railway station building.

10.7 Data Horizon: IT-Based Health Management

Data Horizon in Hiroshima City is in the data health business, analyzing information from health insurance claims and physical check-ups and providing lifestyle disease prevention and health care services. It is notable for adopting the high-risk approach—to prevent a further decline of people who are already ill—instead of the population approach—to prevent disease in the general public. The firm was founded in 1981 and originally it developed ingenious systems for customer management and production management at pig farms and gas stations. Fifteen years ago, CEO Utsumi Yoshio heard from a former official at the MHLW that the increase in Japan's elderly population and lifestyle-related diseases such as diabetes would push Japan's medical expenses through the roof. This offhand comment was the catalyst for shifting his existing IT expertise to embark on a health management business.

Data health businesses require a large number of nurses but there is no need for a night shift as is necessary for hospital work. There may be up to half a million qualified nurses nationwide who cannot work night shifts due to family circumstances, and so many cannot find work at hospitals although they want to find employment (Harada 2010). Data Horizon matches employers looking for nurses with the conditions that nurses find amenable. In the end, nurses who cannot work night shifts—mostly women—can be paired with good, local employment opportunities.

10.8 Niseko Town, Hokkaido: A Town Luring Overseas Talent

Known for its high-quality powder snow, Niseko Town, Hokkaido is visited every winter by many skiers and snowboard enthusiasts from home and abroad.[10] It has also attracted attention as a town where the population has stopped falling and has been increasing since the 2000s. In 2010, the inflow of people in their fifties and sixties outnumbered those leaving. This change indicates that people moving in following retirement is growing. However, a major cause of the town's population growth is the large number of young people in their late twenties to early forties moving in. Starting around 2004, word-of-mouth about the town's high-quality snow spread worldwide, leading to a surge in foreign skiers. The number of non-Japanese nationals in the town's Basic Foreign Resident Register jumped from 11 in 2004 to 162 in 2014. With 28 Australians, 23 British and 21 Canadians, the number of English-speaking nationals is a new feature as well.

Along with the snow, the "Niseko Rules" also enhanced the town's reputation abroad as a ski resort. In the past, skiers would go off-piste to ski on powder snow that had not been compacted by other skiers. This caused a series of accidents, such as avalanches. In an attempt to eliminate these accidents, local volunteers roped off the boundaries in 2001 to keep the courses clear. It was decided that access outside the course was only to be via several gates that were opened when the local Niseko Avalanche Survey deemed it safe. The number of accidents then steadily decreased. These efforts established a perception of Niseko as a "safe ski resort where you can enjoy powder snow." It was featured in international media outlets such as the *New York Times* in 2012, encouraging visits from a growing number of skiers and snowboarders from Europe and the United States (Ondine 2012).

What is distinctive about this town is the serious interest local residents have for the safety of skiers. They do not see accidents outside the course as a question of self-responsibility. There is a strong awareness that Niseko is a tourist destination. Constant streams of tourists arrive in the summer as well when there is no snow. Visitors, including foreigners, take advantage of Hokkaido's majestic nature, such as rafting and mountain climbing. The activities of non-Japanese residents have transformed the ski town of Niseko into a year-round outdoor town. One example is Ross Findlay. He brought rafting to Niseko and as a result the number of summer tourists is now on a par with winter numbers. Guides who specialized in winter skiing also acquired guide qualifications for summer outdoor activities and this yearlong tourism had an impact on solidifying work stability. In 2002, the Hokkaido Outdoor Qualification System began—largely through the efforts of Findlay once again—leading to improvements in the quality of guides through the enhancement of test trials and training sessions. In this way, Niseko became well equipped for people working in tourism to settle there. Similar to the winter "Niseko Rules," non-Japanese and Japanese entrepreneurs cooperated with local residents for summer leisure activities, and a local Non-Profit Organization (NPO) drew up a set of rules for protecting nature and accident prevention entitled the "Shiribetsu River Compact."

10.9 Naoshima, Kagawa Prefecture: An Island Where 100 Times the Number of Residents Visit

Naoshima (Kagawa-gun, Kagawa Prefecture), located in the Inland Sea, is now famous as an island of art. In 1992, the CEO of Fukutake Publishing (now Benesse Holdings, Inc.) built Benesse House, combining a museum of contemporary art and a hotel. From this beginning, an underground museum was added in 2004 and the Setouchi International Art Festival began in 2010. In addition, a building that is used as a bathhouse also serves as a work of art, and the works of world-renowned artists such as Ando Tadao, Kusama Yayoi, and Sugimoto Hiroshi dot the island. Every year, more than 100 times the population visit this island of around 3000. Thanks to the island being featured on CNN and in the guidebook *Lonely Planet*, foreigners account for about 15% of all visitors. Having a refinery for copper and other metals, Naoshima did not initially have any particular affinity to art but the tipping point for turning it into an island of art goes back to the "House Project." From 1998, Benesse renovated a 200-year-old wooden house as a work of art. Local citizens, who also participated, developed and shared their interest in the project and the art. Since then, other activities involving them have taken place, including the *Yago* Project, where house names are displayed on stainless steel designed gate plates, and the *Motomura Noren* Project, where each house hangs out a traditional cloth banner suited to its own image. Local residents acted as volunteer guides for the Naoshima Standard Exhibition (2001, 2006) when the entire island served as an art venue. To cope with the surge in tourists, accommodation and dining options have opened up one after another in renovated old houses and the number of young people returning to the island to work has also increased.

10.10 Sabae City, Fukui Prefecture: Residents Change the Town

As pointed out in the section of Chapter 7 on silver democracy, it seems that in the current environment only problems related to the elderly gain political traction. In response, Sabae City is taking efforts to prevent succumbing to this tendency by encouraging younger generations to get more active in politics. As the winds of the Heisei mega mergers (the ongoing large-scale integration of local governments started in 1995) began to be felt, Sabae was also on the brink of being swallowed up by the storm. There was much talk about Fukui Branding and a Twin City Concept for local industry that would develop from a merger with neighboring Fukui City.[11] However, after two referendums and the recall of the pro-merger mayor, the proposal was completely withdrawn. There were those that criticized this as waffling, but it put into practice the idea that "local government is the foundation of democracy." Makino Hyakuo, who became the new mayor in 2004 after the recall, later implemented various measures to ensure that this experience of community participation was not completed in vain. A typical example of his efforts was the Regional Revitalization Planning Contest launched in 2008. Mayor Makino and young members of the town communicated and exchanged their opinions on a blog, renewed seven times by 2014.[12] Students from both inside and outside the city were asked to propose plans about revitalization, which were then considered by the responsible departments to be incorporated into actual administrative proposals.[13]

In fiscal 2013, the Sabae Regional Revitalization Planning Contest was held for the first time. A plan for "having high school girls act as the new public mainstay for information dissemination" was adopted. A female high school section (known as the *Joshi Kosei*, or *JK* Section) was set up to act as a priming agent for involving the many young people indifferent to resident autonomy.[14] This female division not only handles information dissemination, but has also compiled interviews and the opinions of high school students on public works.[15] The city also established the "Citizen Leading Role Ordinance" in 2010. Hailed as a proposal-model citizen-led initiative, it engaged a variety of plans for the private sector, such as the Encounter Support Business Plan, which provides a venue for unmarried men and women to meet, as well as the Rewarding Life Course, which is aimed to help people over 60 years of age find meaningful pastimes, such as social club activities for drawing, sports, and learning with computers.[16] By 2014, the city adopted more than 100 citizen-led projects and it is currently reviewing the roles of the private sector and government. The city initiated a virtuous cycle by adopting grassroots proposals; the sense of autonomy among citizens increased, which encouraged more business proposals, which in turn increased proposals adopted, boosting the sense of autonomy yet again. Entrusting projects to the private sector also improved efficiency, and the desired effect is to relieve the pressure from a declining birthrate/aging population and financial difficulties.

10.11 Toyama City: Chosen Town Development

Mayor Mori Masashi of Toyama City refers to himself as being "like an entrepreneur," and since taking office in 2002 he has worked to improve the appeal of the region with the keyword "chosen town development." Toyama City statistics show that since 2008, the number of people moving to the city has outstripped the number moving out for five consecutive years (Toyama City n.d.). Toyama City has strengthened public transportation so that people can live there even without a car. In addition, by paying subsidies to residents who move nearer to public transport lines, the city is promoting its transition to a "compact city," where people gather in the center and do not live in outlying areas. Jarred by a sense of crisis that under those conditions Toyama would disappear in three decades, Mori himself attended resident briefings and asked for their input and cooperation to combine city services and condense urban development into a smaller zone. The number of briefings ran as high as 120.

The problem of Japan's endless urban sprawl was detailed in Chapter 6 but the "compact city" can be thought of as a way of dealing with this problem by enhancing public transportation for vulnerable transport users—the elderly, students, the handicapped and people without cars. It is not possible to create a town that truly attracts people unless all residents, starting with such socially vulnerable groups like the elderly and single mothers, find it to be a safe living environment. Mayor Mori has been quoted as saying, "If you can create a community with a certain sense of security, young people will stay and new residents will arrive. I want to create a sense of security about living in Toyama by providing generous support to single mothers, households on welfare, and children's homes, and so on."[17] Based on the idea of seven female staffers at the local city government, a Single Parent Support Program was launched in 2015 for a large number of single-parent households.[18] Specific measures include the city paying its own child support allowance on top of the existing child allowance and medical expense subsidies. Residents have responded by saying, "I can live here even in 20 years' time as the town has a friendly environment to citizens."[19] Although there are still many high school graduates who move away from Toyama today, creating a city that is good for the socially vulnerable to live in will likely lead to the return of residents who have left as well as to an influx of new residents.

10.12 Ritsumeikan Asia Pacific University: Oita's International City

Surrounded by the famous Beppu Hot Springs in Oita Prefecture, Asia Pacific University (APU) was founded in 2000. Of the total project cost of 29.7 billion yen, Oita Prefecture subsidized 15 billion yen, and Beppu City also injected 4.2 billion yen and transferred free of charge 42 hectares of city-owned land. Currently, there are some 5700 students from about 80 countries and regions studying there. As shown in Chapter 8, the most reliable way of attracting overseas talent is through exchange students. The APU-related population including the faculty is about 6000, which almost exactly supplements the 2.24% decline in Beppu's population from 1995 to 2010, which was low for a medium-sized municipality. The intake of young people every year also contributes to the revitalization of the local municipality, as well as that of the nation as a whole by attracting international students then sending them out highly skilled to the Tokyo metropolitan area and elsewhere. APU's goal is to become a venue for knowledge creation aimed at common human goals, bringing together young people responsible for the future from countries and regions around the world, to live and learn together, and understand each other's cultures and customs (APU 2000). Beppu's efforts to pursue education aimed at internationalizing the city instead of the conventional, locally grounded efforts is unique and will have a ripple effect on the surrounding areas.[20]

As a result, APU graduates have a high employment rate with companies seeking employees with a global background,[21] and they play an active part in many regions of the world. These features and reputation attract many young people to the city each year. APU also has a significant effect on both the economic and population fronts, with the university-related population spending approximately 12.1 billion yen a year in Beppu City.

10.13 Fukuoka City: The Women's Choice

Since the Masuda Reports in May 2014 pointed out the risk of local municipalities becoming extinct due to an extreme drop in women of childbearing age, the outflow of young women has attracted a great deal of attention in Japan. Ways to keep women from leaving is a matter under discussion, but there are voices of dissatisfaction since *keep ... from leaving* gives the impression of constraining the self-determination of women.

Rather than merely trying to *keep* women, women must *choose* a city; Fukuoka City is one such example. Many women in their twenties have chosen to live in Fukuoka City compared to other regional central cities. For example, in 2010, for every man in their twenties the proportion of women was 1.09 in Fukuoka, 1.07 in Sapporo, 1.06 in Hiroshima, and 1.03 in Sendai; in particular, the number of females in their late twenties is 15% higher than the commensurate male population in Fukuoka. Conceivably, the major reason for this is the large number of employment opportunities. At 53.9%, the female labor participation rate (the ratio of workers to the total 15-years-or-older female population) is higher by more than two percentage points compared to other regional central cities. This gap can also be seen in the large number of positions in industries such as healthcare and retail where it is easy to create jobs for women. There are many businesses in the fashion-related retail industry—1.71 times that of Sapporo, 1.39 times that of Hiroshima, and 1.32 times that of Sendai. As Chapter 2 noted in this regard, this sense of opportunity is in direct contrast with the Greater Nagoya area where, although the manufacturing sector is expected to stay strong and see a population influx in the future, there is concern an exodus of women will result from the lack of jobs for them. Fukuoka City Hall is also working to create a better working environment for women, such as setting the numerical goal of promoting women to more than 20% of section chiefs or more senior positions in 2018.

10.14 Summary: Leveraging Private Sector Power for the Falling Population Issue

What is important is that the ideas and power of the business sector will be a major driver to solve the population issue. As described above, businesses that produce attractive jobs in the regions will attract people and potentially slow the pace of population decline. Strong companies that survive while maintaining close ties with the region bring vitality to local economies stricken by demography. Of course, many strong companies in a given region will make competition too severe to survive. It is more realistic to foster an exceptionally strong company that acts as a model case or goal for the region to act as a leader. The role of local government in selecting companies should be limited, and public involvement or intervention is not desirable because it prevents normal competition. Government should be devoted to monitoring and infrastructure development, not to helping the weak. It is also important that a number of strong companies cluster together. This means that important business information will be amassed and thereby increases efficiency. Interaction between companies will be put into motion, and highly skilled people will accumulate. Living and educational environments grow better in areas like this, which in turn makes it easier for people to congregate.

Expectation for innovation belongs in the regions. This is because free thinking and new challenges are more possible there than in Tokyo where large, tightly organized companies abound and competition is severe. A parallel is that technological innovation in the United States takes place in Silicon Valley rather than New York City. Today, when there is growing momentum to allow a variety of new initiatives and to support them, is the chance to create innovation. The hurdles to setting up business have dropped significantly. Above all, young people are no longer afraid of entrepreneurship. Furthermore, along with fostering *new* businesses, revitalizing *existing* local businesses is also important. As can be seen from the case studies described in this chapter, innovation is happening in a variety of firms, not only venture companies. There are many opportunities for innovation in traditional local industries such as agriculture, tourism, and commerce. Nor is it impossible to build strong local firms by boldly integrating industries that can generate many jobs such as distribution and retail, restaurants, and tourism.

Every regional company needs to seriously consider pursuing its own economic development using their respective resources and areas of expertise as well as introducing the best practices discussed in this chapter. At the same time, they should be able to make a "strong company" by modernizing management and consolidating governance. It will be strong companies in each region relentlessly pursuing profits that will support the multi-dimensional growth of Japan. This chapter has shown that truly strong companies not only pursue profits, but also invest the revenue generated by their businesses into social contribution activities such as solving community problems, preparing infrastructure, cultural promotion, fostering emerging companies, and others. While "social contribution" tends to recall charitable activities and volunteers, when it comes from the private sector it tends to last longer if they are firmly cemented in the firm's core business. Companies that support the future of Japan need to be just as skillful in *using* profits as they are in *making* profits. With more companies like this in the various regions of Japan, the fuller the life of individuals will be and the more choices they will have. This would lead to a brighter outlook for the future and an increase in the number of women bearing children. We need to face up to the fact that it is only with an economy and the survival of the regions that a rich society can exist in Japan.

After the collapse of the bubble economy, the private sector paid dearly, streamlining its operations and creating sound management, which has resulted in a large pool of retained earnings. The great management challenge now is how to use these savings for the future of Japan. Strong companies have led the way investing their own management resources for the future of the regions and the nation. This trend needs to be built up into a larger groundswell. In the face of Japan's difficult problems like the falling population issue, it is hoped that the government will provide infrastructure and structural reforms so that strong companies throughout the nation can tackle these issues in concert. Although creating "rewarding work with a decent income" and "socially significant, growing businesses" requires an alliance between politics, government, and academia, the leading role falls to the private sector. The ultimate countermeasure against population decline is to involve as many private companies as possible in undertaking such efforts.

Notes

1. Small and Medium Enterprises Agency 2014.
2. Ishikawa Prefecture 2013.
3. "Sixtiary industry" is a Japanese term referring to a combination of agricultural production with primary, secondary, and tertiary industries, including processing, services, retails and agritourism, and creates new value using local resources effectively and efficiently. More information can be found in Ministry of Agriculture, Forestry and Fisheries (MAFF 2010). Information about this school may be obtained from the school website in Japanese (http://www.6ji.jp/6ji/info.html).

4. Fidea Holdings, Ltd. 2009.
5. Nitta, Kaichi, personal communication, January 2014.
6. "A modern home guardian" is about area management that Shimizu created and advocated. Detail can be found in Socio design note website (http://socio-design.jp/post/21008381914/014). The article is in Japanese.
7. The company was launched in April 2012. Detail can be found at the company website in Japanese (http://www.yamorisha.com/outline).
8. MHLW 2013.
9. Dr. Wakatsuki first used this word, "Medicopolis," in 1988. The first usage of the word is cited by "Workshop on medical systems" (2006).
10. Niseko Town 2013.
11. Branding Fukui was an initiative to take the eyeglasses and lacquerware that was being sold under the "Sabae" brand after a municipal merger and increase their overseas sales under the "Fukui" brand. The mayor promoting this believed that it would be easier to sell Sabae products worldwide under the Fukui brand. The Twin City Concept was the brainchild of the Sabae mayor, who at the time was promoting a merger with the city of Fukui. The idea was to make Fukui as the information center and Sabae as the production center, the two cores of the post-merger city. It was propounded by the mayor, who was pushing for the merger. See also *The first major heisei municipal merger* by Tsuji (2007).
12. Takebe, Miki, CEO of NPO L-Community: Takebe returned from Tokyo, created an NPO to promote and support Sabae Town in October 2012. This is recorded in a blog by L community in Japanese (http://www.l-community.com/).
13. Many proposals have made it to the real world as they have been adopted or been incorporated into existing projects. For example, in order to enliven the Sabae Train Station, a vacancy on the second floor of the station building was used to launch a library with a café in 2015.
14. *JK* stands for *joshi kosei* (female high school student). Its website is available in Japanese (http://sabae-jk.jp/).
15. For example, they continuously engage in the planning and execution of the "*Pika Pika* Plan," in which participants pick up trash while enjoying themselves by dressing up in costumes. They also came up with the idea of an app to search open seats in libraries, real-time, jointly developed it with jig.jp, a local IT firm, and provided it to the public. The city website provides detail in Japanese (http://www.city.sabae.fukui.jp/pageview.html?id=14528).
16. For example, social club activities such as drawing, sports, and computers are provided. The city website provides more information in Japanese (http://www.city.sabae.fukui.jp/pageview.html?id=17493).

17. Mori, Masashi, personal communication, Summer 2014.
18. Ibid.
19. A man in his fifties, who had moved from the city outskirts to the center quoted in the *Asahi Shimbun*, December 23, 2009, morning edition.
20. APU 2015.
21. The employment rate was 91% in 2014. For reference, see APU job offer statistics (APU 2016).

References

APU. (2000). Declaration on the occasion of the opening of APU. http://en.apu.ac.jp/home/about/content5/? accessed on February 27, 2017.

APU. (2015). APU job placement statistics for the 2014 academic year. http://en.apu.ac.jp/careers/page/content0148.html, accessed on September 23, 2016.

APU. (2016). Major employers and graduate schools recruiting APU students. http://en.apu.ac.jp/home/career/content7, accessed on September 23, 2016.

Fidea Holdings, Ltd. (2009). Interim report, September 2009. http://www.fidea.co.jp/pdf/h2109.pdf, accessed on September 23, 2016 (in Japanese).

Harada, Hiroko. (2010). Building the foundation for a work-life balance (Waku raifu baransu no jitsugen ni hitsuyona kiban zukuriwo). *Medical Publishing* No. 2905, November 22. https://www.igaku-shoin.co.jp/paperDetail.do?id=PA02905_03, accessed on September 23, 2016 (in Japanese).

Ishikawa Prefecture. (2013). Comprehensive agricultural partnership conclusion with Komatsu (Komatsu tono nogyo ni kansuru houkatsuteki renkei kyoutei teiketsushiki) (Press Release, June 17). http://www.pref.ishikawa.lg.jp/kensei/koho/photogallery/2013/06/20130617/index.html, accessed on September 23, 2016 (in Japanese).

MAFF. (2010). Agriculture towards sixtiary industry (6ji sangyoka ni tsuite). http://www.maff.go.jp/j/shokusan/sanki/6jika/houritu/pdf/6jika.pdf, accessed on September 23, 2016 (in Japanese).

MHLW. (2013). Working group meeting on medical specialists (Aratana senmon'inikansuru shikumini tsuite). http://www.mhlw.go.jp/stf/shingi/2r985200000300ju-att/2r985200000300lb.pdf, accessed on September 23, 2016 (in Japanese).

Niseko Town. (2013). Niseko by the numbers, April 2013 (Suji de miru Niseko, 2013 nen 5 gatsu). http://www.town.niseko.lg.jp/machitsukuri/数字でみるニセコ(H24.5観光抜粋).pdf, accessed on September 23, 2016 (in Japanese).

Ondine, Cohane. (2012, January 12). Niseko, Japan's own St. Moritz. *New York Times*. http://www.nytimes.com/2012/01/15/travel/niseko-japans-own-st-moritz.html, accessed on September 23, 2016.

Small and Medium Enterprises Agency. (2014). 2014 White paper on small and medium enterprises in Japan: fight song for micro business. http://www.chusho.meti.go.jp/pamflet/hakusyo/H26/download/2014hakusho_eng.pdf, accessed on September 23, 2016.

Toyama City. (n.d.). Population trend statistics (Jinko no ido). http://www.city.toyama.toyama.jp/kikakukanribu/johotokeika/tokei/jinkosetai/jinkosetai.html, accessed on September 23, 2016 (in Japanese).

Tsuji, Kayoemon. (2007). *The first major heisei municipal merger: the ordinariness of Kayoemon, the mayor who was recalled* (*Heisei daigappei dai-ichigo Kayoemon kaishoku shicho no nichijosei*). Tokyo: Bungeisha (in Japanese).

CHAPTER 11

Policy Proposals

11.1 Policies

11.1.1 Birth Rate Target

1. The government should create a demographic policy that will both *mitigate* and *adapt* to population decline.
2. Based on the desired ideal of 2.4 children per family, set as a target the achievement of a total fertility rate of 2.1 between 2040 and 2050, and then formulate and implement the policies necessary to reach the goal.
3. Transfer income from households that are not raising children to those that are.

While raising pension entitlements depending on the number of children and the childrearing period, the total pension amount should be reduced by 5% through taxes on the pension income as soon as possible, and the benefit age should be raised to 70 years of age within 10 years.

(Intergenerational assistance has already begun with a macro-economic slide for pensions, which are up 0.9% in fiscal 2015).

4. Amend the medical care system for the elderly in the latter stage of life (the Act on Securing Medical Care for the Elderly), and make the medical burden of the elderly a uniform 30%.

Y. Funabashi (ed.), *Japan's Population Implosion*,
https://doi.org/10.1007/978-981-10-4983-5_11

(When the government created its policy of free medical care for the aged in 1973, both an aging population and population decline were unexpected. In a subsequent reform of the health system for elders, a 10% self-pay burden of medical expense was imposed, which was further raised to 20%. Both changes respectively took about ten years to implement.)

5. Promote measures not only for longevity but also for extending a healthy life expectancy.[1] The retirement system should be abolished, and age-free labor practices and legal systems coordinated so all are able to work to the best of their ability and responsibility regardless of age.

(This can be expected to contribute to improving the balance between work and home for child-raising households by allowing the elderly in good health who wish to continue working to do so, as well as keeping elderly people healthy and not requiring care.)

(The Organization for Economic Co-operation and Development (OECD) report "Live longer, Work longer" predicts that the proportion of inactive elders in OECD countries as a whole will rise from 38% in 2000 to 70% in 2050 if work patterns and the retirement age remain unchanged.[2])

11.1.2 Declining Birth Rate Measures

6. Raise the GDP ratio of family relationship spending—the public funding scale of cash and taxes, childcare leave, and childcare—to the same 3% as in other countries where countermeasures for declining birth rates are successful (e.g., France, Sweden). The GDP ratio of family related spending in Japan was 1.35% in fiscal 2011.

Specifically, financial resources derived from reviewing measures for the elderly should be invested in declining birth rate measures.

The Basic Law on Declining Birth Rate Society Countermeasures should be amended and implementation of this goal made compulsory for the government.

7. Lower the relative poverty rate and the child poverty rate to the OECD average.

Specifically, the relative poverty rate of 16.0% should be lowered to 11.3%, and the child poverty rate from 15.7% to 13.3%.[3] We should note that the poverty rate of children in single-parent families is more than 50%, and requires immediate action.

(Given the tightness of public finances, political handouts should be avoided and income restrictions for social security benefits and other expenditures should be stricter.)

8. Switch to a social system that prioritizes pregnancy, childbirth, and childrearing.

   (a) Create in-house childcare facilities in offices and factories, and a work environment that watches over children.
   (b) Create a family-friendly working environment by introducing flex-time systems for workers commuting with children.
   (c) Introduce railway fare incentives for children during summer vacation.
   (d) Create safe places in the community where children up to junior high school age can learn and play, helping to both foster the child's social skills and the parent's work–life balance.

9. Promote matchmaking within the private sector level for unmarried men and women.

(Around 50% of unmarried men and women of 25–34 years of age cite "not meeting an appropriate partner" as their reason for not marrying. Arranged marriages once accounted for nearly 70% of all marriages but have fallen to 5% in recent years.)

10. Give all children equal rights.

Eliminate legal categories of discrimination between the so-called legitimate and illegitimate children and create a system where it is easier to get married.

11. Promote national initiatives linking foster parenting and special adoptions, so that all babies born have the chance for a happy future.

This would require strengthening consultation and support systems, as well as the budgets and staffing for key actors and child guidance centers. The government should ensure children's welfare and smooth administrative responses—and that these are delivered without disparity to all regions—and should also consider enacting legislation in the future to promote these efforts.

(Official statistics place the number of abortions in Japan at around 200,000 per year, but the actual number is believed to be higher. While there are various background factors, most likely some of these are unplanned or unwanted pregnancies. Rebuilding a more effective and wide-ranging special adoption system—currently used only in a limited way—might induce some women with an unplanned pregnancy to choose to give birth. The so-called Aichi Scheme that has been both ambitious and successful with adoptions is a useful point of reference.)[4]

11.1.3 Work–Life Balance

12. Fundamentally change the life course of young people, and especially women. Women in their 20s and early 30s should be encouraged to position this time as a dual life stage for education, marriage, childbirth, child-raising, and work, and the 40s as a full-fledged working life stage.
13. To realize this life course strategy, work hours should be shortened, overtime abolished, and telecommuting promoted for those in their 20s and early 30s. There should also be a substantial overhaul of corporate career development plans for this age group.
14. After a mother has given birth, the government should send out a "Father's Handbook" outlining information such as the respective rights and obligations of parent and child, childcare leave, and a child-raising planner.
15. Establish a social security system that enables labor flexibility.

    (a) Amend the social security system, which assumes a system of lifetime employment and a gender division of labor, to build more neutral social plans for choosing work styles.

(In Japan, poor labor flexibility means that new graduates are unevenly distributed to certain popular industries; this distortion then becomes fixed over the long term, which makes reforming or transforming Japan's industrial structure difficult. Labor flexibility is desirable for optimizing human resource allocation in line with changes in society.)

- Make equal pay for equal work an overarching principle.

(For labor with the same relative value in terms of quantity and quality, the principle of equal pay regardless of gender, age, and race would eliminate disparities especially based on gender and work status.)

- Promote an expanded application of employee insurance and return the national pension system to its original purpose of being for the self-employed.

(It is possible to correct disparities in social insurance by expanding and applying it to all employees above a certain wage income, 58,000 yen or more/month, as well as to end dependent spouses.)

16. Build a social system better able to take advantage of skilled elders.

(E.g., science education/experiment guidance, schoolchild care/junior and senior high school student after-school tutoring, individual learning guidance for children with development disabilities by retirees with the required expertise.)

11.1.4 Education

17. Expand a scholarship system that does not require repayment.

Steep tuition fees for higher education results in the excessive reliance of young people on their parents and delays the independence of youth. To overcome this, scholarship programs that do not require loans should be expanded.

Keeping in mind the case of European universities, measures should be taken such as doubling national university corporation operating grants and private school subsidies.

18. Offer greater opportunities for pre-school education.

(Not only higher education, but also pre-school education breaks the poverty chain.)[5]

19. Adopt "recurrent education" to improve career paths by offering higher education opportunities after raising children.[6]

11.1.5 National Land Policy and Population Relocation

20. As well as creating a national land policy with a vision for regional population relocation, link that vision to specific measures and budgets.
21. Lower the number of people who cannot get access to basic services less than an hour from their homes, and aim to make this number as close to zero as possible.

This will require the expansion of social infrastructure, ensuring public transport services, relocation of residents through city aggregation, as well as the enhancement and diversification of home services.

(Basic services cover those generally supplied to a population on the order of several thousand to 10,000 people, such as basic education [elementary and junior high schools], the postal service, and general clinic functions. The government should regularly survey and analyze the status quo and closely monitor developments to ensure the necessary services are not lacking.)

22. When the supply costs of administrative services and facilities to residential areas with a low population density are significantly high, the government should be allowed to lower the quality and quantity of services.
23. The number of regional metropolitan areas lacking the required advanced urban facilities for maintaining a working-age population should be reduced to zero in order to restore and maintain the population density.

(Advanced urban facilities cover those services generally supplied to a population in the hundreds of thousands, such as shopping centers or department stores, cinemas, museums, universities, and law firms. The government should publish data on the status quo.)

24. Build regional comprehensive care and compact cities centering on cities with a high likelihood of maintaining a population in excess of 100,000 by 2050.

11.1.6 Land and Housing

25. Create a society where use takes precedence over ownership.

(During high growth and economic expansion, people aspired to own land and a home. This is because they were able to accumulate wealth through reselling or renting. Such real estate appreciation cannot be expected in a low-growth society with a declining population. As the meaning of ownership fades, a move from ownership to use has begun in some areas. A declining population should prioritize those who use rather than own, and greater levies should be imposed on people who own but do not utilize the property.)

26. Switch housing tax benefits from those who buy new homes to those who acquire existing homes, especially the acquisition and renovation of vacant houses.
27. Public housing should prioritize collective houses.[7]

11.1.7 Metropolitan Area Measures

28. Tokyo must enhance its resilience to natural disasters in order to increase sustainability.

(According to estimates, the risk of natural disaster risk in Tokyo and Yokohama is eight times higher than Osaka-Kobe-Kyoto)

29. Establish legal requirements for non-sound fire alarm equipment with the aim of a 100% penetration rate for the increasing number of elders, care recipients, and people with disabilities in the metropolitan area.
30. Companies should jointly develop a business continuity system for an entire metropolitan area so that its social and economic functions are not disrupted even in times of crisis.
31. Reduce to zero by 2035, the number of buildings designed and built with pre-1981 seismic standards, or those that have not been subsequently earthquake reinforced.
32. Raise Tokyo's total fertility rate to the national average.

(Tokyo was 1.13 in 2013, compared to a national birth rate of 1.43. If Tokyo's birth rate can be improved, it would also help resolve the problem of the Tokyo metropolitan area pulling in the populations of other prefectures, and in the long term it would also be an advantage for responding to aging in the city.)

11.1.8 "Strong Cities" and "Strong Companies" in the Regions

33. Aim at creating 50,000 jobs for young people each year through the growth of core companies in the region.
34. Create *strong companies* and *strong cities* in the regions to attract back young locals that have moved away, and encourage other new residents. Business innovation is created on the back of the private sector.

11.1.9 Overseas Workers

35. Develop employment and career paths for foreign students following completion of their studies as a framework leading to future settlement in Japan. Specifically, promote initiatives for the expansion of corporate hiring schemes for international students, the adoption of enhanced post-hire training programs, and respect the specific skills of international students.
36. Implement autumn enrollment at universities to attract high-caliber foreign students.
37. Urgently create a framework for accepting 150,000 foreign workers annually. Aim for a 5% population ratio of overseas talent by 2050. If 150,000 overseas workers are accepted every year from 2015, this will become 5.25 million people by 2050, or approximately 5% of the total estimated population of 97 million. *(The proportion of foreign-born citizens within the total population is 15.5% in Sweden and 11.9% in France.)*
38. Increase tourism from overseas, creating a situation where there always are many foreign workers in Japan. Being an "open state" would become the norm for Japan, and would also bring economic benefits equal to the annual consumption of one million Japanese residents.[8]

11.2 Creating a Policy Environment

39. Demographic policy is a long-term national project. It requires the formation of sustained political will and a broad national consensus.
40. Create a "Next Generation Ministry"

(The population problem has a depth and range that cannot be covered by a single Ministry of Health, Labour and Welfare (MHLW). Nor can a coor-

dinating body like the Cabinet Office cope alone. It would be better to create an administrative organization that oversees the entire population problem with legislative, budgetary, planning and analytical functions. The Democratic Party of Japan once advocated the establishment of a "Child and Family Ministry." However, the ministry proposed here would be comprehensively responsible for population movement related to the social mobility and relocation policies (national land policy is also relevant), rather than limited to marriage-related increases in population, birth and childcare support policy. In addition, a specialized demographic research and analytical institution should also be housed in this ministry comparable to the National Population Institute of France, which was founded in 1945.)

41. Establish the "One Hundred Year Demographic Research Council," reporting directly to the prime minister, to advise on and monitor demographic policy.

(Changes in government are likely to become the norm in the future. If so, it will be necessary to create a policy coordination mechanism between the ruling and opposition parties to avoid swings in population policy with each change of government and to coordinate national preparedness and a consensus on population policy. As a mechanism to guarantee this, a bipartisan National Congress would be established consisting of 13 people, a chair along with six men and six women. One member of each gender should represent each generation by decade—from those in their twenties to those in their seventies. The Research Council would be composed of experts from the National Assembly, the government and the private sector. It would monitor, examine and make recommendations concerning the government's implementation of demographic policy every five years. The Council would run for a century. Each member's term of office would be ten years. The baton would be passed every ten years; watching, giving orders, supporting to see the effects of population policy ascending in a hundred-year long policy marathon.)

42. Government updating of an Internet "Population Notebook" and "Population Clock" indicating the current status of Japan's population on a daily basis.

(A national discussion on population policy is not possible without the provision of government data. To avoid repeating the mistakes of the MHLW's independent prediction of a third baby boom, demographic data should be

widely published in order to contribute to research by private think tanks and academic researchers. Information disclosure also includes data related to life infrastructure not only demographic data itself. For example, the Ministry of Land, Infrastructure, Transport and Tourism (MLIT) should publicize the date on the per capita cost of road maintenance, because this is essential when considering rationalization, restructuring and regional alliances for life infrastructure due to a falling population.)

43. Establish an automatic mechanism for correcting disparities in the one-vote system.

(To ensure that Japan does not head in the direction of a "tyranny by majority" in a silver democracy, one-vote disparities that are structured around regions with many elderly voters should be corrected.)

Notes

1. The difference between average life expectancy and healthy life expectancy is the "unhealthy period" where daily life is limited. The difference between the average life expectancy and healthy life expectancy (no period of limited daily life) was 9.13 years for men and 12.68 years for women in 2010 (MHLW 2013, 2015).
2. OECD 2006.
3. The relative poverty rate refers to the percentage of household members who do not meet the poverty line. The poverty line is half the amount of median equivalent disposable income.
4. A program where newborns are placed directly into foster care from the hospital (MHLW 2011).
5. Science Council of Japan 2014.
6. Recurrent education refers to switching repeatedly between full-time employment and full-time study in order to acquire necessary professional knowledge and skills. It began in Sweden in the late 1960s.
7. A rational joint-living residence where each resident has his/her own space as well as shared space, and families live independently but also enjoy broad ties with other inhabitants not necessarily related to them.
8. The number of foreign tourists visiting Japan in 2013 was 10.36 million. When converted to tourist consumption, the annual spending per capita of the residential population (1.24 million yen) is equivalent to ten tourists (Japan Tourism Agency 2014).

References

Japan Tourism Agency, MLIT. (2014). Tourism measure (Kanko ni kansuru torikumi ni tsuite). http://www.kantei.go.jp/jp/singi/keizaisaisei/jjkaigou/dai8/siryou1.pdf., accessed on October 14, 2016. (in Japanese).

MHLW. (2011). Guideline for fostering children (Satooya itaku gaidorain). http://www.mhlw.go.jp/stf/shingi/2r98520000018h6g-att/2r98520000018hlp.pdf., accessed on October 19, 2016. (in Japanese).

MHLW. (2013). Reference materials for implementing basic policy on improving nutrition and eating habits, and health promotion by administrative dietitians in regions (Chiikini okeru gyosei eiyoshiniyoru kanrieiyoshi niyoru kenko zukuri oyobi eiyo shokuseikatsu no kaizen no kihon shishin wo jissen surutameno siryoshu). http://www.mhlw.go.jp/bunya/kenkou/dl/chiiki-gyousei_03_zentai.pdf., accessed on October 14, 2016. (in Japanese).

MHLW. (2015). Looking at life expectancy and healthy life expectancy (Heikin jumyo to kenko jumyo wo miru). http://www.mhlw.go.jp/bunya/kenkou/dl/chiiki-gyousei_03_02.pdf., accessed on October 14, 2016. (in Japanese).

OECD. (2006). Live longer, Work longer. https://www.oecd.org/employment/livelongerworklonger.htm., accessed on March 3, 2017.

Science Council of Japan. (2014). Proposal for a healthy next generation (Sukoyakana jisedai ikuseini kansuru teigen). http://www.scj.go.jp/ja/info/kohyo/pdf/kohyo-22-t195-2.pdf., accessed on October 19, 2016. (in Japanese).

Index[1]

[1]Note: Page number followed by 'n' refers to notes.

Y. Funabashi (ed.), *Japan's Population Implosion*,
https://doi.org/10.1007/978-981-10-4983-5

The manufacturer's authorised representative in the EU is Springer Nature Customer Service Centre GmbH, Europaplatz 3, 69115 Heidelberg, Germany. If you have any concerns regarding our products, please contact ProductSafety@springernature.com

Printed and bound by CPI Group (UK) Ltd, Croydon, CR0 4YY

30/09/2026

02225447-0001